THE GENESIS ENIGMA

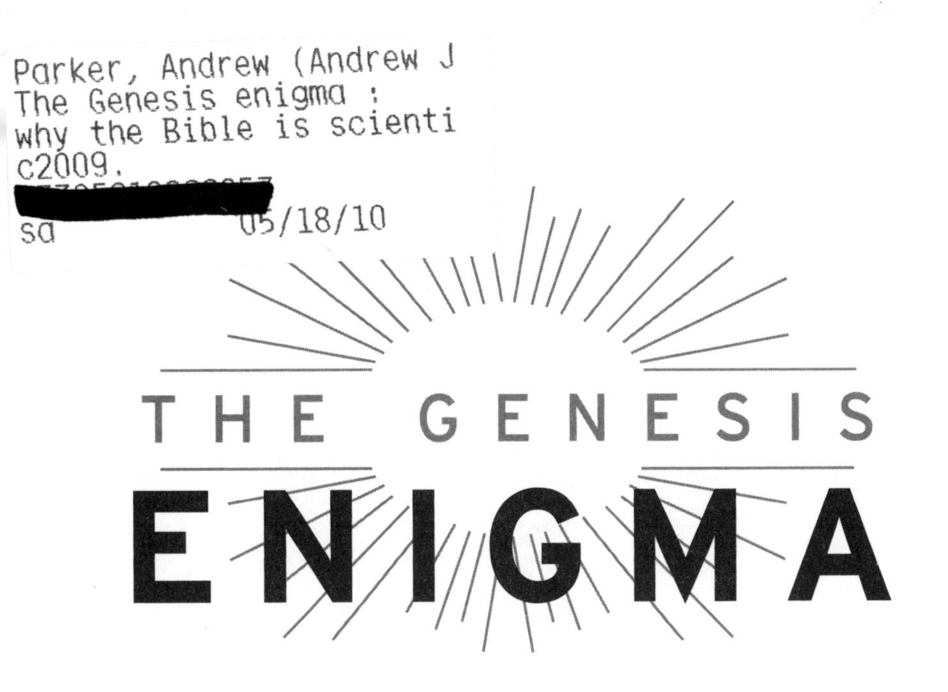

THE GENESIS ENIGMA

WHY THE BIBLE IS SCIENTIFICALLY ACCURATE

ANDREW PARKER

DUTTON

DUTTON
Published by Penguin Group (USA) Inc.
375 Hudson Street, New York, New York 10014, U.S.A.
Penguin Group (Canada), 90 Eglinton Avenue East, Suite 700, Toronto, Ontario M4P 2Y3, Canada (a division of Pearson Penguin Canada Inc.); Penguin Books Ltd, 80 Strand, London WC2R 0RL, England; Penguin Ireland, 25 St Stephen's Green, Dublin 2, Ireland (a division of Penguin Books Ltd); Penguin Group (Australia), 250 Camberwell Road, Camberwell, Victoria 3124, Australia (a division of Pearson Australia Group Pty Ltd); Penguin Books India Pvt Ltd, 11 Community Centre, Panchsheel Park, New Delhi—110 017, India; Penguin Group (NZ), 67 Apollo Drive, Rosedale, North Shore 0632, New Zealand (a division of Pearson New Zealand Ltd); Penguin Books (South Africa) (Pty) Ltd, 24 Sturdee Avenue, Rosebank, Johannesburg 2196, South Africa

Penguin Books Ltd, Registered Offices: 80 Strand, London WC2R 0RL, England

Published by Dutton, a member of Penguin Group (USA) Inc.

First printing, October 2009
1 3 5 7 9 10 8 6 4 2

LIBRARY OF CONGRESS CATALOGING-IN-PUBLICATION DATA
Parker, Andrew (Andrew J.)
The Genesis enigma: why the Bible is scientifically accurate / Andrew Parker.
p. cm.
Includes bibliographical references and index.
ISBN 978-0-525-95124-7 (hardcover)
1. Bible and science. 2. Cosmogony. 3. Cosmology. 4. Creationism. 5. Bible. O.T. Genesis—Criticism, interpretation, etc. I. Title.
BS651.P275 2009
231.7'65—dc22 2009020782

Printed in the United States of America
Set in Granjon
Designed by Leonard Telesca

TO MY SON, LORENZO

CONTENTS

PREFACE

It has been said many times, by rational people who believe in God, that the truth of the stories in the book of Genesis is not to be measured by their agreement with the facts of modern astronomy and biology. Rather, it is argued, their truth is bound up with their effectiveness in explaining for the Old Testament writers the origin and destiny of the world and humanity in the light of their belief in God. In other words, the creation account at the beginning of Genesis helped the ancient Israelites who produced it to cope with and understand their world. In some way, this story of how they and their world came about satisfied their inner consciousness. But no one in more recent times would ever have considered that this creation account could be scientifically accurate, which is why modern theologians have felt compelled to arm themselves for debate with the rationalization above.

Things may be about to change.

In this book I will be arguing that the latest understanding of how the world and all life on it came to develop and evolve, as demonstrated by solid, evidence-based science, reflects exactly the order of events as set out in Genesis. It is the culmination of a most extraordinary and unanticipated chain of events that began in 1990 at the Australian Museum in Sydney, where I worked under the auspices of Drs. Jim Lowry, Noel Tait, Penny Berents, and Pat Hutchings. I had set out to describe a whole new community of crustaceans known as seed shrimps, but soon became distracted by—literally—a flash of green light. The seed shrimps were producing iridescence, but according to the literature they were not supposed to. Investigating further, I found that the bright color resulted from diffraction gratings on the animals' antennae. Diffraction gratings consist of very fine parallel grooves in a surface; they are what cause, for instance, the color seen in holograms on credit cards. They were well known to physicists, but hitherto unknown in nature. This itself became another green light to look for more—and I found lots more, thanks to my job at the museum, where I spent several years observing all the major groups of animals represented in tropical seas.

The authority on seed shrimps was Professor Lou Kornicker at the Smithsonian Institution in Washington, D.C. While working with Lou in 1994 I also saw some of the prize possessions of the Smithsonian, the fossils of the Burgess Shale (more about these in chapter 6). The Burgess Shale fossils had become famous for revealing the Big Bang of evolution—an event known as the "Cambrian explosion"—of around 520 million years ago, long before the dinosaurs, when life was exclusively marine. I noticed, in black-and-white reproductions of these long-extinct animals, fine parallel lines drawn on some of their spines and shields. These lines themselves were too coarse to be diffraction gratings; but could they, I wondered, be vestiges of such a feature?

I was given permission to examine the valuable fossils under high-powered microscopes, and in three species I did indeed find diffraction gratings. These animals would have sparkled with bright colors when they lived all that time ago. But why? The answer to that lay in the reason why some animals display bright colors today—to send visual messages. Eyes existed in the Burgess Shale animals too, so the visual world and the natural "arms race" between prey and predator, each with endlessly evolving innovations to give it the edge over the other, could be extended back to at least their era. But how much farther did they extend?

This question led me to search for the first eye to evolve on earth. I found that it had evolved in a group of trilobites, which lived around 521 million years ago. That's just before the Cambrian explosion began. It was not hard to put two and two together.

In 1998 I published my "light switch theory" in a scientific journal, *Proceedings of the Royal Society of London*. This theory holds that it was the sudden introduction of vision in life forms on earth that triggered the Cambrian explosion. Although initially lost in a sea of alternative theories, today the light switch theory not only still holds strong but has become the "last man standing" in the critiques of experts in the field (such as in a 2007 review by Harvard professor Charles Marshall).

A couple of years later, when I moved to Oxford University's Department of Zoology (thanks to Professor Marian Dawkins), I wrote a full exposition of the light switch theory. The result was a popular science book called *In the Blink of an Eye: How Vision Sparked the Big Bang of Evolution*, published on both sides of the Atlantic in early 2003.

I received numerous letters from readers of this book suggesting something completely unforeseen—that there were parallels between the scientifically generated history of life on earth and the creation account given in the first chapter of Genesis. It was the evolution of vision—the cornerstone of the light switch theory—that had caught their attention, to decipher the ever-puzzling biblical phrase "Let there be lights . . . to divide the day from the night" (verse 14). This appears after the initial "Let there be light" or formation of the sun (verse 3). The light switch theory became the final number in a combination lock guarding a biblical secret. Mainly out of courtesy, I gave the matter brief attention. Needless to say, as a scientist I would never entertain the "seven-day" creation story, according to which the universe and all life were supposedly created literally in seven actual days, or other such irrational ideas. So, I wondered, what's all this about scientific accuracy in the book of Genesis?

Without expecting to find anything, I discovered a whole series of parallels between the creation story on the Bible's first page and the modern, scientific account of life's history. This at least made me think. The congruence was almost exact and it had assembled by accident—it was free of intervention by human hands. Most importantly, I was considering the scientific account first—a detached review of the facts and chronology of evolution—which ruled out partiality or clouded judgment.

My thoughts developed when I moved from Somerville College, Oxford, to Green College (now Green Templeton College), later in 2003. At Green College I was fortunate to have the chance to talk to Professor John Lennox over many a lunch and coffee. John is a mathematician, but one with an ardent interest in, and extraordinary knowledge of, religious doctrine and history. He is also well versed in the "science versus religion" debate, which at the time was becoming more topical than ever. John introduced me to his world, and I was surprised to discover within religion some good sense and logic. He also introduced me to some of his friends from the religious arena, such as professors Alister McGrath and Alan Millard, who left me with much thinking to do. John drew the little religion within me to the tip of my tongue and I decided to look into this Genesis Enigma in more detail. The word "enigma" arose often during these conversations—we could say Genesis Enigma was born here.

I drew further inspiration from seeing for myself some of the discoveries made in the Holy Land at the British Museum, introduced to me by Jonathan Tubb, who had unearthed some of the artifacts himself. I marveled at the finds and correspondence of Leonard Woolley and T. E. Lawrence ("Lawrence of Arabia") at the Palestine Exploration Fund's headquarters in London. The defining moment, nevertheless, was probably when I *saw* the Genesis Enigma while looking up at Michelangelo's masterpiece on the ceiling of the Sistine Chapel.

In many ways, the central aim of *The Genesis Enigma* is to represent the voice of those people who wrote to me about the implications for religion veiled within *In the Blink of an Eye*. As I said, there are, inexplicably, a whole series of parallels between the Genesis account of creation and the history of our planet and of life as told by science. Chapters 2 to 8 of *The Genesis Enigma* will explain each parallel in detail—for the more detail is examined, the more convincing and remarkable I believe the parallels become. One question I will be asking in this book is this: Could it be that the creation account on page one of Genesis was written as it is *because* that is how the sequence of events really happened?

If that were true, what would it mean? Since we know the scientific account to be correct through repeatable, proven experiments, it surely follows that the Genesis account must be correct, too. But the astonishing point is that the Genesis account has (as it were) *no right* to be correct. Consider the identity of its writer—when and where he (or they) lived; there's no way they could have guessed this thesis as it stands. The best guess available at the time would have been something very different; no one had access to the information, or the techniques, needed to deduce the scientific account. So why—and how—were those words of Genesis written? I do my best to make some sense of this question in the final chapter of this book.

In the first chapter I will be looking at the evidence that the Old Testament is a work of nonfiction. Looking at it historically and archaeologically, what evidence is there to corroborate the events described in this part of the Bible? The remaining chapters will explain how we came to develop our scientific understanding of the formation of the universe and diversity of life. This will help to convey the magnitude of the Genesis Enigma—how incredible a correct guess at the current science from a "blank slate" of knowledge would be.

The first five books of the Bible (the Pentateuch) as we know them today are central texts to the Christian and Jewish religions, but are also relevant to Islam. The Genesis Enigma concerns the Pentateuch specifically, and so may be interesting to those of all three "Abrahamic" faiths.

I would like to finish this preface with two thoughts. The first is Erasmus Darwin's idea that the cause of causes is God, not the causes themselves. This allows us to accept all of science along with a God who perhaps made the most basic building blocks of the universe, from which everything self-assembled via the laws of physics and biology. The second thought is that of Lord Robert Winston, who observes that spirituality does exist, and that the question to be asked is whether spirituality is God-given or whether it evolved at a moment when humans needed it in order to equip individuals to get through an unfavorable period in geological time. Myself, I will go away and reflect further on these ideas. In the meantime it would be a great shame if my findings were either misused in an attempt to suggest that scientists themselves are unsure about science, or pounded out of all recognition into support of the seven-day creation premise. If this happens, then *The Genesis Enigma* will have been wrongly interpreted, and my endeavor will have been in vain. That would really be a big step backward.

Certainly I am not at a particular moment in my life where I need the comfort of religion, so I can feel objective rather than "faith-based." But to accept purpose in the universe, and consequently the existence of God, is inevitably to take on an element of faith. If only there were reliable evidence for such purpose, to relieve faith of the entire burden . . . I hope that the reader will find some in *The Genesis Enigma*.

THE FIRST BOOKE of Moſes, called Geneſis.

CHAP. I.

1 The creation of Heauen and Earth, 3 of the Light, 6 of the Firmament, 9 of the Earth ſeparated from the Waters, 11 and made fruitfull, 14 of the Sunne, Moone, and Starres, 20 of Fiſh, and Foule, 24 of Beaſts and Cattell, 26 of Man in the Image of God. 29 Alſo the appointment of foode.

N * the beginning God created the Heauen, and the Earth.

2 And the earth was without forme, and voyd, and darkeneſſe was vpon the face of the deepe: and the Spirit of God moued vpon the face of the waters.

3 And God ſaid, *Let there be light: and there was light.

4 And God ſawe the light, that it was good: and God diuided † the light from the darkneſſe.

5 And God called the light, Day, and the darkeneſſe hee called Night: † and the euening and the morning were the firſt day.

6 ¶ And God ſaid, * Let there be a † firmament in the midſt of the waters: and let it diuide the waters from the waters.

7 And God made the firmament; and diuided the waters, which were vnder the firmament, from the waters, which were aboue the firmament: and it was ſo.

8 And God called the * firmament, Heauen: and the euening and the morning were the ſecond day.

9 ¶ And God ſaid, * Let the waters vnder the heauen be gathered together vnto one place, and let the drie land appeare: and it was ſo.

10 And God called the drie land, Earth, and the gathering together of the waters called hee, Seas: and God ſaw that it was good.

11 And God ſaid, Let the earth bring foorth † graſſe, the herbe yeelding ſeede, and the fruit tree yeelding fruit after his kinde, whoſe ſeede is in it ſelfe, vpon the earth: and it was ſo.

12 And the earth brought foorth graſſe, and herbe yeelding ſeede after his kinde, and the tree yeelding fruit, whoſe ſeed was in it ſelfe, after his kinde: and God ſaw that it was good.

13 And the euening and the morning were the third day.

14 ¶ And God ſaid, Let there bee * lights in the firmament of the heauen, to diuide † the day from the night, and let them be for ſignes and for ſeaſons, and for dayes and yeeres.

15 And let them be for lights in the firmament of the heauen, to giue light vpon the earth: and it was ſo.

16 And God made two great lights: the greater light † to rule the day, and the leſſer light to rule the night: he made the ſtarres alſo.

17 And God ſet them in the firmament of the heauen, to giue light vpon the earth:

18 And to * rule ouer the day, and ouer the night, and to diuide the light from the darkneſſe: and God ſaw that it was good.

19 And the euening and the morning were the fourth day.

20 And God ſaid, *Let the waters bring forth abundantly the ‖ mouing creature that hath † life, and foule that may flie aboue the earth in the † open firmament of heauen.

21 And God created great whales, and euery liuing creature that moueth, which the waters brought forth abundantly after their kinde, and euery winged foule after his kind: and God ſaw that it was good.

22 And God bleſſed them, ſaying, * Be fruitfull, and multiply, and fill the waters in the Seas, and let foule multiply in the earth.

23 And the euening and the morning were the fift day.

24 ¶ And God ſaid, Let the earth bring forth the liuing creature after his kinde, cattell, and creeping thing, and beaſt of the earth after his kinde: and it was ſo.

25 And God made the beaſt of the earth after his kinde, and cattell after their kind, and euery thing that creepeth vpon the earth, after his kinde: and God ſaw that it was good.

26 ¶ And God ſaid, *Let vs make man in our Image, after our likeneſſe: and let them haue dominion ouer the fiſh of the ſea, and ouer the foule of the ayre, and ouer the cattell, and ouer all the earth, and ouer euery creeping thing that creepeth vpon the earth.

27 So God created man in his owne Image, in the Image of God created hee him; * male and female created hee them.

28 And God bleſſed them, and God ſaid vnto them, * Be fruitful, and multiplie, and repleniſh the earth, and ſubdue it, and haue dominion ouer the fiſh of the ſea, and ouer the foule of the aire, and ouer euery liuing thing that † moueth vpon the earth.

29 ¶ And God ſaid, Behold, I haue giuen you euery herb † bearing ſeed, which is vpon the face of al the earth, & euery tree, in the which is the fruit of a tree yeelding ſeed, * to you it ſhalbe for meat:

30 And

INTRODUCTION

What Is the Genesis Enigma?

Why is the Bible's creation story written as it is?

The first page of Genesis features no humans at all. The substance of heaven and earth—and earth's earliest creatures—arise spontaneously at God's command. Light is mentioned twice, once in "Let there be light" and then in "Let there be lights . . . to divide the day from the night." Early life is specifically associated with the sea. Vegetation appears after "Let there be light," but before day is separated from night.

Any study of the Bible will show that the rest of the Old Testament is written with scrupulous care through prolonged, meticulously recorded narratives of human adventure. Yet the description of the creation as it appears in the Bible seems to make no sense.

Could it be that the creation story in Genesis was written as it was, complete with its seemingly odd order, because that is in fact the correct order of events at the beginning of the world? Modern science has revealed in stunning detail how the world and all of the life on it came into being. We are now in a position to make a comparison of the two accounts.

Central to *The Genesis Enigma* will be our modern understanding of the history of life on earth, a small part of which, called the "light switch theory," is my work. I explained in the preface how publication of my book *In the Blink of an Eye* prompted a number of people to write to me suggesting that there were parallels between the scientifically proven history of life on earth and the creation account in Genesis, and how, when I investigated this further, to my surprise I found a whole series of parallels between the creation story in Genesis and the proven, scientific account of life's history. The box on page xxiv sets out a basic summary of these parallels, using the 1613 edition of the Authorized Version of the Bible—the "King James version." The facts have not been skewed in order to contrive them: They emerge unsought and, most importantly, *the science was considered first*. This point cannot be overemphasized, and it removes all bias.

So the light switch theory and the letters I received after the publication of *In the Blink of an Eye* had got me thinking; but the inspiration for the *Genesis Enigma* came from a separate starting point: Rome. It was seeing Michelangelo's great painting—the world's first (and perhaps most successful) attempt to visually realize the Old Testament's creation story—on the ceiling of the Sistine Chapel in the Vatican that made me think about how creation *really* happened. Was the version Michelangelo showed us a true one? It was while looking up at that glorious ceiling that I first wondered what the ancient writers of the Bible knew about the beginnings of everything—and how that knowledge was understood at the time Michelangelo painted his masterpiece. Could they have got it right?

Eventually it struck me that a biblical enigma exists that is on the one hand so cryptic it has remained camouflaged for millennia, and on the other so obvious one cannot miss it. All in all we have not seen the woods for the trees. What's more, the enigma has been revealed on some of the most sacred bricks and mortar in the Christian world—the ceiling of the Sistine Chapel. Just like creation's pride of place in the Bible, it has been there in one of the world's most famous works of art, directly above the heads of popes for centuries. Michelangelo painted the creation story as one would expect someone without scientific knowledge to represent it—using the human form. That way everyone could identify with it. But the writer of Genesis opted instead for a cryptic, more abstract description. Why?

The Sistine Chapel was designed to match the exact internal dimensions of the Temple of Solomon in Jerusalem—twice as long as it is high—as specifically detailed in the Bible. Work began in 1477 with construction of the cannonball-proof walls, ten feet thick at their base. At the top of these walls ran a walkway, with arrow slits for archers and holes through which to pour boiling oil on attackers below. Offering protection for cardinals, the chapel also housed soldiers, and later prisoners, in the rooms above the vault.

The chapel's last stone was laid in 1480, as the finest artists in Italy, including Sandro Botticelli, were already heading for Rome to decorate its walls. Even the Florentine magnate Lorenzo de' Medici ("Lorenzo the Magnificent") had granted his painters leave of absence to undertake this special task. They were commissioned to record in fresco scenes from the lives of Moses on one side and Jesus on the other. The ceiling itself lay lower in the priorities of the incumbent pope, Sixtus IV, who chose for this supposedly unworkable surface a standard design involving gold stars laid upon a blue background. Twenty years later another pope, Julius II, summoned Michelangelo with instructions to repaint the ceiling, presenting the artist with a plan for a Romanesque design, a geometric pattern with interconnecting circles, squares, and hexagons. But Michelangelo had other ideas. He reverted to his favorite subject—the human body.

We know that Michelangelo often frequented the Sistine Chapel to soak up the ambience while the scaffold for his own work was being constructed. As he looked up to where the workmen were hammering away, he also saw the process whereby the stars in the existing sky scene were becoming . . . nothing. Empty white plaster had replaced the depictions of solid, celestial objects. Stars had been painted on nothingness; now they were reverting to nothingness. He was witnessing the creation in reverse. Michelangelo chose for his subject something totally original: Not just the story of Adam and Eve, Noah, and the usual characters who had traditionally adorned classical religious art, but the very beginnings of the creation story, how the earth itself and life came into existence. It was a subject entirely new to art, possibly because it had always eluded human understanding.

We know that Michelangelo settled upon these biblical events: *God separating light from darkness*, *God creating the planets*, and *God separating the sky and water*. In the Chapel ceiling's first panel an anthropomorphic God prises pale, swirling clouds from their shadowy counterpart, suggestive of the beginning of something. God's powerful physique is visible through his billowing robes, swirling in the wake of his motion. A sun and a moon are evident in the next panel, although God in his commanding human form once more dominates the scene, this time accompanied by the further human-like forms of cherubs. In the third panel God—the all-powerful and all-wise creator in person—divides two shades of blue with his path. Thus the artist brought the power and vitality of sculpture into the realm of painting.

On October 31, 1512, four years after work had commenced, the newly frescoed Sistine Chapel ceiling was finally unveiled. The pontiff, Julius II, observed Michelangelo's tour de force "with immense satisfaction," commented Giorgio Vasari in his *Lives of the Artists*. "When the work was thrown open, the whole world could be heard running up to see it, and, indeed, it was such as to make everyone astonished and dumb."

The sight of the Lord God in full length—complete with fingers and toes, caught in mid-flight—was completely new. *This* is how humans wanted to see their God. This is the very image we often still form when told of a creator; it is an image that one can relate to and that holds some meaning.

Nevertheless, the writer of the Genesis creation account up to the stage before the creation of man had not employed the image of a humanlike God. This is in sharp contrast to the writer's use of the human form in the remainder of his biblical work, from which it is clear that he fully appreciated the power of anthropomorphic forms when communicating unknown entities. But Genesis 1:1–25 was without human reference.

That struck me forcefully. On that ceiling was Michelangelo's attempt to understand, interpret, and realize the creation story. And that is exactly what we will try to do in this book—but, rather than using the mediums of paint and plaster, our tools will be those of modern science.

At the beginning of this introduction we saw a preview of the solution to the Genesis Enigma in the suggestion that, astonishing as it seems, the Genesis creation account was written as it is because that is how the sequence of events really happened. But it is not until the scientific details are added that it becomes how clear how remarkable the parallels between the biblical and scientific accounts actually are.

WHAT IS THE GENESIS ENIGMA?

Genesis	***The scientific history of earth***
And God said, "Let there be light"	The formation of the sun (around 5,000 million years ago)
And God said, "Let the waters under the heaven be gathered together unto one place, and let the dryland appear"	The formation of the seas and the separation of land areas (around 4,200 million years ago)
"Let the earth bring forth grass, the herb yielding seed [vegetation]"	The beginnings of life, including single-celled, photosynthetic organisms ("plants")(around 3,900 million years ago)
"Let there be lights . . . to divide the day from night"	The first image-forming eye evolved and the visual information used. The lights were turned on for animal behavior and evolution (around 521 million years ago)
"Let the waters bring forth abundantly the moving creature that hath life"	The Cambrian explosion—evolution's Big Bang (beginning around 520 million years ago). Life was exclusively marine at this time. All the animal phyla that exist today evolved their characteristic forms at this period. These facts are not common knowledge. Only an experienced biologist would know this.
God created the great whales, and every living creature that moveth, which the waters brought forth abundantly after their kind . . .	Life flourished and large animals evolved, from sharks to giant squidlike forms. But again, this took place in the sea, as implied here in Genesis. Later in life's history, animals conquered the land and the diversity reached new heights.
. . . and every winged fowl after his kind	All animals adapted to the vision of predators, except for birds, which could afford not to because they can escape predation through flight, and so generally can avoid camouflage colors. It is fascinating that sea creatures and birds should be singled out. These are, respectively, the main characters and exceptions in life's history book.

Our scientific understanding of the history of the earth and of life has been developing for centuries and this understanding, coupled with the stories of scientific endurance that lie behind it, will be threaded into *The Genesis Enigma*. Intertwined with it will be the history of the earth and of life as provided by the book of Genesis, so that the comparable stages in science and biblical text will lie side by side.

As I hope will become apparent, the latest discoveries of modern science paint a fascinating and vivid picture of the earth's (and life's) earliest stages of existence—and do so in persuasive detail, much more so than even Michelangelo's Sistine ceiling. With the benefit of this scientific perspective we now understand much more about how the universe and life came to be than our ancestors could possibly have known. This raises some of the questions central to the Genesis Enigma. As noted in the preface, the Genesis account, written when it was, has no right to be correct, so can it really be true? And if so, how? And who could have written such a prescient description?

But before we look at how the science and Genesis compare, let's take a quick look at the Bible from a historical point of view. We must first ask the question: Why should we consider the Old Testament to be a reliable source at all?

CHAPTER 1

TRUTH

The Old Testament as Factual Record

Gothic light, mahogany furniture, fine bindings. The exorbitant prices in one of London's leading antiquarian bookshops seemed justified, given that many of the books here were the first copies to leave the printing presses as they set forth into the world—in some cases, to change the world. Today the touch, sight, and smell of a first edition act as a sensory time machine. I hoped that the one I was looking for would take me to some sandy streets three millennia ago.

I was here for the volumes on the lower, reinforced shelves—the Bibles. There were lots of them, dating back to the sixteenth century, written in English, Hebrew, Latin, and Greek. Unlike my local junk shop, the feeling here was that every page on this shelf had already been thoroughly researched: There was no dust to blow off, no new discoveries to be made. I got sucked into a patched-up tome bound in shades of tan, which despite the ravages of time retained its grandeur. This was one of the earliest King James Bibles, a folio edition from 1613, typeset in the antiquated font known as

"English blackface." It is the first page of this text from which I took the phrases that featured in the introduction to this book. But how did the words I gazed upon come to take their place on that page? They were supposed to be the words of God, but I was staring at black-ink letters on antique rag paper. There was no apparent connection.

The purpose of this chapter is to examine the reliability, or otherwise, of the Bible as a historical source. Since this book has sat on the lower shelves of bookshops, and on our tables, for so many centuries, frustrating all our efforts to look around and beyond it, surely such an exploration must be warranted.

There are two approaches we can take to probe the Bible for testable facts. To begin with, the focus will fall on the origin of the biblical prose, since, unlike most other books on the shelves, the Bible lacks an author's name on its spine and a date of origin on its copyright page. The tales may be familiar to us, but where do they come from? Is this book to be considered fact or fiction? In this chapter I want to look at the Old Testament as a complete map containing all those chronicles, fully scrolled out so that any patterns will show themselves. Then the spotlight can shift to the colossal question of truth in the Bible—whether the text is an accurate representation of real events, or a deliberate fabrication. If we are to believe in the sacred messages in the biblical text, it would help us to know that they are supported by sturdy pillars of truth. Some form of independent evidence—evidence from beyond the Bible's texts themselves—would help. If patterns do emerge in the Old Testament, it may be appropriate to apply nonbiblical evidence at least to test their authenticity. If tests prove positive, we will have found truth in the Bible. For the purposes of this book we can consider truth as "what really happened," or factual accuracy.

From pillars of truth to *Pillars of Wisdom*—just a few shelves west of the bibles lay the T. E. Lawrence classic. And there's a connection. The legendary Lawrence of Arabia was once an archaeologist, and at one point in his career he set off on an expedition to the Middle East with his colleague Leonard Woolley "to identify sites mentioned in the Bible . . . and to study the neighborhood of Ain Kadeis, supposed to be the Kadesh-Barnea of the Israelite wanderings." It was not this enterprise, of course, that made him famous; but his legendary endurance in the desert also led to the exposure of biblical evidence. To understand what Lawrence and Woolley were searching for, one needs some knowledge of the contents of the Old Testament and how they came together.

The ancient Israelites believed that, through some means of divination or channeling, God "spoke" directly to Moses and the other prophets (men such as Samuel, Nathan, Gad, Isaiah, Haggai, and Zechariah)—respected figures who stood at the very center of public life. The prophets spoke their messages to the crowds, and as significant public figures they would have been listened to keenly; their words would have been honored, and they would also have been recorded. What we can test is whether these people really existed and when. The evidence (historical and archaeological) and the academic study of the Bible begin here, at the moment when the prophets' messages were first written down.

The Making of the Bible

Cuneiform is the oldest form of writing known to us. The technique, which uses marks made by pressing a wedge-shaped tool (usually a cut reed) into moist clay, first appeared in the Middle East in 3200 BCE, lasting up to the second century CE. By the early second millennium BCE, cuneiform had spread all over the biblical regions and was widely used. The point here is that writing was in place in ancient Israel by the time of the prophets of the Old Testament—the first and early second millennia BCE.

The idea of the prophets' messages passing through generations orally, suffering the adverse, distorting effects of a game of telephone, is at least partly fictional. Indeed, in ancient Israel writing was recognized as durable and reliable and was evidently well appreciated. In the Bible itself, of the judgment on Amalek in the book of Exodus, someone took the trouble to record that God said to Moses, "Write this for a memorial in a book." Further, the cuneiform tablets were properly archived and protected. Eventually, some centuries later, the sacred words of prophets that had been recorded on clay tablets passed through processes of compilation and editing before becoming figuratively set in stone.

As religion spread in ancient Israel, separate archives of sacred literature would have begun to grow in different religious centers (there are indications of this in the Old Testament). Ultimately this sacred literature was gathered from diverse locations into a single collection in Jerusalem. In this process the biblical writings would have encountered editors of some geographical and genetic diversity.

In the learned center of Jerusalem further editing would have taken place, which may explain why many Old Testament books became anonymous. So long as this editing was carried out by other prophets, the sanctity of the literature would not have been affected. Then, at some point, the writing and editing of these sacred texts must have ceased and the disparate collection must have been compiled into a single volume, which we know as the Old Testament. That point in time, which we can work out from the events described in the latest book in the Old Testament and from future events known to us that were not recorded there, can be placed in the fourth century BCE.

Today, we refer to the accepted books of the Bible as "canonical" or as part of the biblical canon. What this means is that, over time, scholars have come to accept certain of the ancient Israelites' religious books as authentic, inspired books that contain the Word of God, while rejecting other books as lacking that divine inspiration. The Old Testament accepted as canonical today by Christians consists of thirty-nine books; the Jewish Bible contains only twenty-four.

Thus we can identify three stages in the generation of the Old Testament—formation (or composition), redaction (or editing), and selection (or recognition as canonical). At the end of this process the collection as a whole was invested with supreme public authority. The whole process of Old Testament production, beginning with the first words written in cuneiform, lasted about one thousand years.

The oldest copies of the Old Testament that we have today were written in Hebrew—the language of the Jews—with some Aramaic (also a Semitic language, from the kingdom of "Aram" or Damascus). Before the time of Jesus, a Greek-speaking colony of Jews at Alexandria translated the Old Testament into the Greek language; the resulting text was known as the Septuagint, from the Latin for seventy, since it was supposed to have been the work of seventy-two translators. This was finished by the second century BCE. The oldest complete copy in existence today is the Codex Sinaiticus, which was written in the mid-fourth century CE and preserved in the monastery of St. Catherine in Sinai, Egypt, close by the mountain where Moses is said to have received the Ten Commandments. At present, parts of the Codex Sinaiticus are also held in the British Library, the library of Leipzig University in Germany, and at the National Library of Russia, although the monks at Sinai are not at all happy about this.

Today the standard Hebrew version of the Old Testament is the Masoretic text. Although this dates back only to around 1010 CE, the strict copying techniques of the Masoretes—groups of Jewish scribes and Bible scholars working primarily in Israel between the seventh and eleventh centuries CE—have resulted in the preservation of the authoritative texts that existed in Christ's time. Indeed, the word "Masoretic" comes from the Hebrew word *masoreth*, meaning traditional.

The Masoretes, believing themselves guardians of the sacred text, exercised extreme, bordering on obsessive, care in making new copies of their Hebrew Bible. All parchment used had to derive from ceremonially clean animal skins and had to be tied together by strings taken from similarly clean animals. Their ink had to conform to a special recipe. They could write nothing from memory, and each word copied from an authentic version of the text had to be read aloud first. Each scribe had to wipe his pen before writing the word for "God" (Elohim), and wash his whole body before writing the word for "Lord" (Jehovah). Strict rules governed the forms of letters, spaces between letters, words and sections, the use of the pen, the color of the parchment, and so on. Each complete scroll of parchment had to be proofread by three other Masoretic scholars within thirty days of its completion, and one mistake on a sheet condemned that entire sheet. Finally, every word and letter would have been counted, and if a letter had been omitted, or an extra letter inserted, or if two letters touched each other, the entire parchment would again be condemned and destroyed at once. Jewish copyists continue this tradition today.

An older version of the Hebrew text was discovered in 1947 in certain caves along the northwest coast of the Dead Sea. These "Dead Sea Scrolls" date back to between around 250 BCE and 135 CE. It is important to note that when the Masoretic text, the Dead Sea Scrolls, and the Septuagint are compared, their religious doctrines agree (though outside the Masoretic text there are inevitable errors in copying and variations in spelling). This provides an excellent test of the authenticity of the Masoretic Old Testament. But how do we go from here to the King James version—the version we are using in this book?

In the first two centuries CE, Jesus' apostles and other missionaries began to spread the Gospel into other lands, and many people embraced the new Abrahamic religion. They needed copies of the Old Testament in their own tongues, and so it was translated into Armenian, Syriac, Coptic, Arabic, and Ethiopic. Across the Roman colonies of Africa, Latin was the language most widely understood, and so around 150 CE the first Latin edition appeared, soon followed by an improved edition for the Christians in Italy. These were in turn superseded by the grand "Vulgate" (*editio vulgata*)—the common, that is, current or accepted—version made by St. Jerome in the fourth century CE by request of Pope Damascus in 382. St. Jerome, a monk and the most learned scholar of his day, made his first translation of the Old Testament from the Greek Septuagint, and then revised it working from the original Hebrew version. St. Jerome's Vulgate, incorporating both Old and New Testaments, has remained the official text within the Catholic Church to this day, aside from some tinkering under Pope Sixtus V in 1590 and Pope Clement VIII in 1593. The English Douai version of the Old Testament (published between 1582 and 1610) is a translation of the Vulgate, and is the foundation on which nearly all English Catholic versions are still based. But King James VI of Scotland, who took the English throne as James I in 1603, was Protestant, and England was by this time a Protestant country. Though himself personally fairly tolerant in matters of religious faith, James's Protestant allegiance was hardened by the Gunpowder Plot in 1605, in which Guy Fawkes and other Roman Catholic conspirators attempted to blow up the Houses of Parliament, and as part of his campaign to strengthen the reformed church in England he ordered a new translation of the Bible in English. The Protestant translators of what would become known as the "Authorized" or "King James" version of the Bible were considered to have been chosen by God, in the manner of the prophets and apostles of old who penned the original scriptures. They were selected on two criteria: devotion to the faith, and linguistic ability. Around forty-seven such devout scholars were involved in the translation, aided by many hundreds of Protestant ministers and believing linguists throughout Britain. They were governed by six committees, two of which sat at Westminster, two at Cambridge, and two at Oxford, and all of which were presided over by Bishop Andrewes, a master of languages (including Hebrew). After each committee of translators had finished the part assigned to it, the translation was subjected to the criticism of the remaining five committees. Finally, a further committee reviewed the complete work.

It has been said of James I's translators that they were scholars of the highest order, and that few if any of today's scholars can compare in their understanding of the original languages, let alone in their faith and commitment to the truth above tradition. Although I wouldn't want to speculate on the truth of that assertion, it is clear that they believed the text they were handling was the very Word of God. *They had absolutely no doubt in their minds that the account of creation given in Genesis was true*.

So that's how the Bible I hold in my hands today came to be. Now, can we perform a rigorous, even scientific analysis in an attempt to reveal some truth in the Old Testament? If so, where on the pages should one look to find a testable hypothesis?

The Bible overflows with *real* dates, descriptions of *actual* events, and biographies of *real* people (though admittedly perhaps less so in the early parts). Importantly, the history component of the Bible is something we can test against independent, nonbiblical sources for accuracy. Here is our first chance to test whether the Bible is a reliable source.

As a history book, the Old Testament chronicles the kings and peoples relevant to Israel. The Old Testament is also a book of law, in matters of both faith and conduct. History and law are described in relation to acts of God—or rather, the history and prophetic interpretation of the acts of God are told in relation to Israel. Such setting of universal issues in a familiar context is hardly surprising. Critics reporting on a play today, for example, base their comments on the world they know, and make comparisons with their personal experiences.

Taking this analogy a little further, other critics watching the same play will undoubtedly provide conflicting reviews, because *different experiences and different environments will have shaped their opinions:* Consequently they will have different expectations of, and different responses to, the play. This point is also relevant to the compilation of the Bible, a process that took place over many hundreds of years, carried out by people with disparate preconceptions and allegiances, some of whom reported on stories already treated by others. This may explain why stories that are duplicated in the Old Testament differ markedly from one version to another.

Now we turn to examine whether the history recorded in the Old Testament, in the people immortalized, events recorded, and places on the map, actually existed. It's time to turn to T. E. Lawrence's foray into the desert along with his colleague Leonard Woolley, a journey that uncovered real evidence for Old Testament sites. Conveniently, an account of this journey also serves to provide a flavor of how men lived and thought in the Holy Land in more recent times and, because the desert sun has preserved (i.e., there's only one way to live in these extreme conditions) an ancient way of life, probably in biblical times too. As the nineteenth-century paintings of David Roberts reveal a more sparsely populated Holy Land of the days before mass tourism—one perhaps closer to life in biblical times than to the twenty-first century—so the literary images painted by Woolley and Lawrence may offer us useful insights. In the background of these images lie the local environments—and we should pay some attention to these too, to get a feel for the desert ourselves.

Lawrence, and Woolley, of Arabia

Born in 1880, Sir (Charles) Leonard Woolley was the son of a clergyman and was expected to follow in his father's footsteps. However, the call of archaeology lured him away, and after graduating from Oxford University he took up a position as assistant keeper of Oxford's Ashmolean Museum. He began working with Lawrence in 1912, in Egypt.

Thomas Edward Lawrence was eight years Woolley's junior. Also educated at Oxford, he studied the crusader castles of Palestine and began his excavation work under the influential D. G. Hogarth in 1911. Within just a few years he developed a rare understanding of Arabia and its inhabitants, thriving in the harsh desert environment. Lawrence's strong pro-Arab allegiance shines through his literary classic *The Seven Pillars of Wisdom*, his account of warfare in Arabia published in 1926.

In 1913 Woolley and Lawrence began archaeological excavations on behalf of the British Museum at Djerabis on the Upper Euphrates river—the supposed site of the ancient Hittite capital of Carchemish. To find out more about their expedition, I made the short trip from the antiquarian bookshop in central London where I had found my King James Bible to rather less salubrious, yet nevertheless curious, surroundings just around the corner, the headquarters of the Palestine Exploration Fund (PEF).

The PEF was once an archaeological heavyweight. Founded in

1865, it was the principal British exploration society in the Holy Land and is the world's oldest established organization for the study of the Levant, or ancient Near East. Today it is a shadow of its former self. The imposing building it once occupied is now leased to wealthier inhabitants, leaving the PEF to make the most of the basement and a room beneath the stairs, equipped with relic furniture from incompatible eras. Still, it has coped well with its decline in fortunes, and the rent allows it to live on. This is good news because, like the affluent book dealer's shop in the same neighborhood, the PEF is a venerable treasure trove.

First on show were the contents of a large mahogany cabinet drawer: the fragments of pottery Leonard Woolley and T. E. Lawrence collected in the Levant. Then there was a complete collection of books written by Woolley and Lawrence, although on these, in contrast to the first editions around the corner, the dust had been allowed to settle. And there was more. I wiped clean a black two-volume folder box from which, astonishingly, the original handwritten and typed letters of Woolley and Lawrence and their associates emerged as individual pages. Some had become fragile, others were a little faded or rain-damaged, but—typical of all the lovingly curated PEF collections—all were punctiliously and systematically arranged in their folder. These letters tell the story of Woolley and Lawrence's archaeological expeditions in the Bible lands, and their subsequent conversion to a military operation. I began to feel the sand between my toes.

At the turn of the nineteenth century into the twentieth, the exploration of Palestine became a race between Britain, France, Germany, and Russia to find biblical evidence. Woolley and Lawrence had gambled on a site known locally as El-Kala'at (within the site known by westerners as Djerabis), a horseshoe-shaped embankment on the River Euphrates, being the Carchemish of old.

Carchemish was an important biblical landmark. This was an ancient city of the Hittites, whose state first developed in Mesopotamia between 2000 and 1700 BCE, and who sold land to Abraham in Genesis. Later, in 605 BCE, Carchemish became the site of an important battle where the Babylonians, under the leadership of Nebuchadnezzar II, defeated the Egyptians. The name Carchemish occurs on Babylonian tablets dating from 2000 BCE, and in an Egyptian reference from 1470 BCE. It disappears from history, however, in the seventh century BCE.

Most archaeologists searching for Carchemish had overlooked Djerabis, but in 1876 the archaeologist George Smith of the British Museum had spotted similarities between the monuments existing at Djerabis and known Hittite stones in other localities. In his diary, at the moment he passed through the site, Smith recorded: "Grand site: vast walls and palace mounds: 8,000 feet round: many sculptures and monoliths with inscriptions: site of Karchemish." Smith died soon after this visit, but his words were enough to set his successors Woolley and Lawrence on their way to the desert.

Fully clad in the English archaeologist's uniform of shirt and trousers, jacket and brogues, Woolley and Lawrence examined for themselves the citadel mound that lay on the bank of the River Euphrates and mapped the outer rampart of the site. They made excavations in areas that taunted with the remains of buildings, combing the ground for evidence and drawing up plans to scale. They photographed numerous statues and inscribed stones, and concluded that most of the remains were Hellenistic or Syro-Roman, dating back only to later settlements of the area—false leads that might lead archaeologists to the wrong conclusions. Even the name Djerabis had no Hittite connection; it had appeared with increasingly modern colonization by Turks and Arabs who had migrated to the region, and was perhaps only three or four centuries old.

Then half a dozen Hittite sculptures and five Hittite slabs were uncovered, and these, along with ancient references to the location of Carchemish, confirmed the site's identity. Egyptian, Assyrian, and Hebrew references placed the city near the Euphrates; one source stated more specifically that it lay to the west, or Syrian side, of the river, while images of the citadel of Carchemish found on the bronze gates of the city of Balawat revealed a location on the very banks of the river. Finally, ancient Egyptian records set it in close proximity to Til-Barsip, modern-day Tell Ahmar. Again, this fitted with Djerabis. Taking all this into account, there was only one site in this region that contained an imposing, pre-Christian citadel and fortifications in the right location, and that was El-Kala'at at Djerabis. The biblical Carchemish had been found.

At the end of 1913 Woolley and Lawrence received a letter from the War Office in London instructing them to suspend their excavations at Carchemish. By now they had acquired an additional asset, more desirable to British military intelligence than their broken pottery: They knew the desert of the Levant like the backs of their hands. They also knew the people, and could advise troops how to gain the tolerance and even cordiality of a tribe. For instance, Lawrence noted of the Bedouin—the nomadic, Arab desert herders in the Middle East—"Their conviction of the truth of their [Islamic] faith, and its share in every act and thought and principle of their daily life is so intimate and intense as to be unconscious, unless aroused by opposition. Their religion is as much a part of nature to them as is sleep or food." Whereas for Christians of western Europe then (and today) religion is a separate, and separable, sphere of life, for the Muslim Bedouin, and for the Jews of Old Testament times, it was an integral part of their existence. Even when we acknowledge this intellectually, it is hard fully to comprehend so radical a difference in outlook; but it is a thought worth keeping in mind as this book progresses.

The War Office removed Lawrence and Woolley from the British Museum's expedition and placed them under the aegis of the PEF. In this capacity they were redirected to the Negev desert (in what is now southern Israel), the deserts of northeastern Sinai, and to the Arabah Valley directly south of the Dead Sea, an area termed the "Wilderness of Zin" in the Bible as adjacent to Kadesh-Barnea, the "holy place of the [Israelites'] desert wanderings" as described in the book of Exodus. Their official business was to conduct an archaeological survey of the area. Their real purpose was to provide cover for British military mapping. At the time, this area was part of the Turkish Ottoman Empire, which was suspected of leaning toward Germany, while Egypt and Sinai were controlled by the British Empire. But despite their new role as intelligence agents, Lawrence and Woolley continued to take their official archaeological business seriously and set themselves the personal goal of locating the Bible's Kadesh-Barnea itself. The Exodus and desert wanderings of the ancient Israelites are a focal point within the Old Testament, and Kadesh-Barnea was their headquarters for thirty-eight of their forty years of discipline (the ancient Israelites were consigned to wander the desert for forty years after they had turned from God). Data to support the desert wanderings would be gold to biblical archaeology.

In 1913 the earliest evidence of Israel's existence outside the Bible had appeared on the Merneptah Stele—writings carved in stone and discovered in 1896 by the British Egyptologist Sir Flinders Petrie at Thebes, in the mortuary temple of Pharaoh Merneptah, who ruled Egypt from about 1213 to 1204 BCE. The stele has been dated to between 1210 and 1207 BCE, and as a result it is accepted that the Israelites emerged as an ethnic group at around this date. But further evidence was desperately needed.

As Lawrence admitted, French, German, American, and English

travelers had already recorded the principal archaeological sites of the Negev in some detail. He and Woolley, however, could contribute something extra to the search. Decisively, they both spoke Arabic with complete fluency and so they were able to breathe life into the existing formal drawings of the biblical evidence. In addition to the known architectural depictions of ruined fortresses and raided tombs, Woolley and Lawrence considered also the broken pots and jugs, envisaging the wine flowing at mealtimes, and from the lives of the ordinary desert dwellers among whom they moved drew a soundtrack of shouting and laughter over a background of crackling fires. An appreciation of the lives and wisdom of the Israelites of old gave Woolley and Lawrence a view through those ancient eyes, and a sense of how they might have negotiated the biblical desert.

It had been suggested that Ain Kadeis (today on the border between Egypt and Israel) was the Kadesh-Barnea of old, although the evidence for this was questionable. Woolley and Lawrence realized that this was a land of nomads, communities constantly on the move, driven by natural conditions and human conflict. People and neighborhoods would have shifted their location throughout the history of this region, and so not only were the names of towns and villages often lost, but even when recovered they could not simply be replaced in their original positions. A village would move when its spring dried up and another water source emerged in a neighboring valley. To use a contemporary spring as a landmark, as others had done in this case, was poor archaeology.

Woolley and Lawrence set out to correct past mistakes. They arrived in Gaza early in January 1914, where they stocked up with stores and tents. From Gaza they followed a track to Beersheba, passing through a wide, undulating plain of deep, rich soil; although it was now void of trees and houses, evidence of an older and more settled civilization existed everywhere. South of Beersheba the landscape changed. The rolling plain gave place to barren hills and ridges and soft limestone, overlain with loose gravel and flint. The general effect of the country was desolate, though the broken terrace walls provided evidence of industrious husbandry in the past.

This sketch of the country and the descriptions in the following paragraphs are based on Woolley and Lawrence's own imagery. You will notice that their painstaking descriptions lack any reference to animals. That's interesting. Over a considerable area, observed while wearing their "ancient Israelite" hats, animals were out of sight and out of mind. The significance of the apparent lack of fauna will become clear . . .

As Woolley and Lawrence approached the ruins of the town of Khalasa, the stony hills were replaced by wide sand dunes, dotted with scrub. The country became flatter and less broken and contained a wide plain of light but good soil, most of which was tilled by the Arabs, whose cluster of tents stood near the well. Woolley and Lawrence's next stop, near the ruins of the town of Esbeita, revealed a contrast—a great belt of drift sand, with ridges 15 meters (or about 50 feet) high. Only the shrubby tamarisk bushes, characteristic of much of North Sinai, could possibly push their way through the sand here. Further along, the sand in turn surrendered to rough limestone hills, on which Woolley and Lawrence noted the bands of flinty strata—a term that indicated some degree of geological understanding on their part. This land had once been a seabed, and sediment falling from the water above and building up on that bed had become rock. The many different layers of sediment formed over time were the strata observed today, after that one-time seabed had been lifted above the water by geological activity.

As the valley between the hills came to an abrupt end, Woolley and Lawrence encountered a vantage point from which to view the great Wady Migrih. Much of this valley was barren, with red

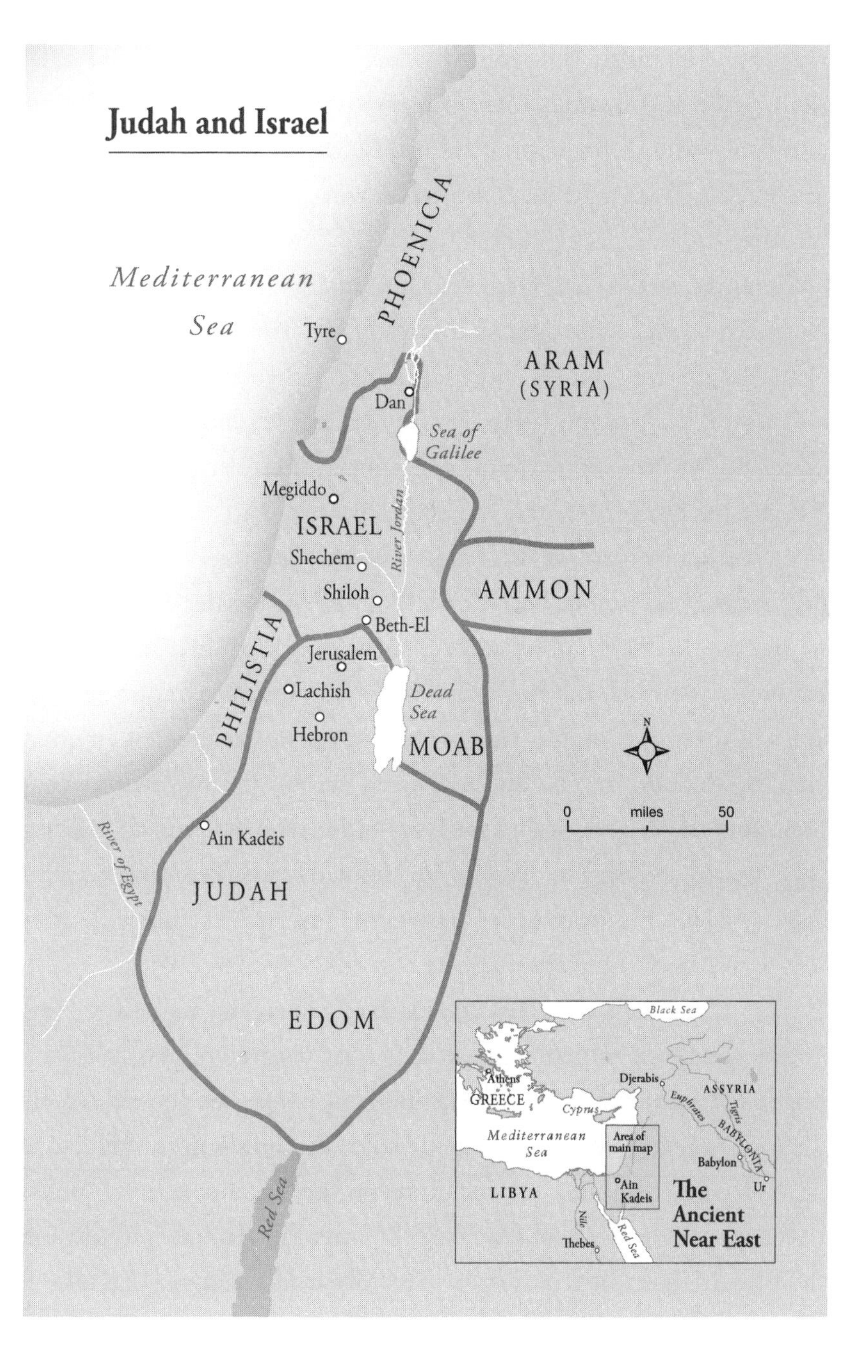
Judah and Israel
Mediterranean Sea
PHOENICIA
Tyre
ARAM (SYRIA)
Dan
Sea of Galilee
Megiddo
ISRAEL
River Jordan
Shechem
Shiloh
AMMON
Beth-El
PHILISTIA
Jerusalem
Lachish
Dead Sea
Hebron
MOAB
N
0 miles 50
Ain Kadeis
River of Egypt
JUDAH
EDOM
Red Sea
Black Sea
Athens
GREECE
Cyprus
Djerabis
ASSYRIA
Euphrates
Tigris
BABYLONIA
Mediterranean Sea
Area of main map
Babylon
Ain Kadeis
LIBYA
Ur
The Ancient Near East
Nile
Thebes
Red Sea

ridges of flint rising between stretches of scrub-dotted sand; but in the middle of the valley a stream gave life to some walled gardens amid the ruins of ancient buildings. By the next valley, the hint of fertile conditions had passed, giving way again to drift sand and scrub. More flora, yet still no fauna.

Scrambling over scree-covered hills against an impressive (albeit barren) mountain landscape, Woolley and Lawrence approached Ain Kadeis toward the end of January 1914. In one valley they encountered a group of five Arab tents surrounding a man-made water hole—more contemporary evidence of an ancient way of life. Then they broke through the foothills that skirt the Ain Kadeis plain. From the town of Ain Kadeis itself, following an examination of the ruins, Lawrence headed south and Woolley returned north.

In their 1915 PEF publication *The Wilderness of Zin*, Woolley and Lawrence describe the last remains of rough terrace walling in the sides of the valley of Ain Kadeis. Here they reported late Christian remains in the form of eight poor ring-graves, some Byzantine pottery, and the foundations of a farmhouse. But other archaeological remains, specifically those of the ancient Israelites, were absent. Nevertheless, the local geography provided a significant clue that this was indeed the Kadesh-Barnea of old.

Stretching westward was the Darb el-Shur, the road of the Israelite forefathers, which suggested an easy return to Egypt. To the north, the same road ran to Hebron, from where the ancient Israelite spies had viewed the Promised Land. To the south ran the main road from Elath, one of the stations on the Exodus route. Finally, to the east lay a choice of roads leading into the Arabah and toward the mountain traditionally known as Mount Hor. In essence, all the roads used by the ancient Israelites were there. Adding this to the general topography of the site, Woolley and Lawrence concluded that surely Ain Kadeis *did* lie on the site of Kadesh-Barnea.

New reinterpretations of Woolley and Lawrence's finds, and indeed those of all early archaeologists, are taking place today under the banner of the new academic subject of Syro-Palestinian archaeology. This reflects a major theoretical and methodological change, a revolution that has ousted the old "biblical archaeology." In particular, the results of investigations are not expected to uphold existing beliefs. The emphasis now is on revealing the evidence as it is unearthed from the ground on a systematic dig, not in hopes of confirming previously held thoughts and expectations. That's not to say that all biblical archaeology was prejudiced. But the new methodologies have been welcomed, since the evidence is recorded as it is found, leaving it open to new interpretations as perspective and the weight of evidence change in the future.

Of particular relevance here is the modern analysis, using radiocarbon dating, conducted on charred organic remains collected from Ain Kadeis by geoarchaeologist Hendrik Bruins in the 1980s. With his archaeologist colleague Johannes van der Plicht, Bruins measured the decayed portion of carbon in these remains of human activity. This accurate method established a date of the second millennium BCE—the time of the Israelites during their forty years of discipline. The recent trend of opinion had begun to move against the idea that people from this era had inhabited Ain Kadeis. But now Woolley and Lawrence were proven right—they *had* located the Kadesh-Barnea of old.

Another fascinating piece of evidence for the historical accuracy of the Bible emerged in 2007. Austrian Syro-Palestinian archaeologist Dr. Michael Jursa is one of the few scholars in the world who can read cuneiform script. Some time after he had begun to reinterpret the British Museum's well-documented Babylonian tablets—long overlooked since their original excavation—a familiar name suddenly emerged.

To the inexpert (which is most of us) the cuneiform tablets may all look the same, but sometimes they contain riches, albeit not immediately obvious riches. Jursa's discovery was the description of a mundane commercial transaction that took place in 595 BCE. A little dull so far, perhaps, but as well as the date, the tablet named an official, one Nebo-Sarsechim. The confirmation that Nebo-Sarsechim had existed as "chief eunuch" in 595 BCE was pure gold. According to chapter 39 of the book of Jeremiah, a character called Nebo-Sarsechim was present with King Nebuchadnezzar II himself at the siege of Jerusalem that is known to have taken place in 587 BCE.

To find a cuneiform reference to someone connected with these remarkable times is rare enough; but evidence from nonbiblical sources for the existence of any individual named in the Bible—other than kings—is fantastically rare.

Even the great Leonard Woolley overlooked the significance of one of his own finds. In September 1922, after two years as a prisoner of war (he had been captured by the Turks during World War I) and Lawrence's retirement from archaeology, Woolley repacked his field bags and, with his assistants, boarded a steamship to the Middle East. He was on his way to Basra in present-day Iraq to lead the joint British Museum and University of Pennsylvania excavations at the ancient site of Ur, the birthplace of Abraham. Ur was also the burial site of many Sumerian royals. The Sumerians emerged in 7000 BCE, and by 3000 BCE they had formed a series of city-states to control the area known as Mesopotamia, between the rivers Tigris and Euphrates. At Ur, Woolley discovered tombs of great material wealth. Inside these tombs were paintings depicting ancient Sumerian culture, which gave another glimpse of life in Old Testament times. His telegram to the University of Pennsylvania—couched in Latin, to protect his news from interception—read: "I found the intact tomb, stone built and vaulted with bricks, of Queen Puabi adorned with a dress in which gems, flower crown, and animal figures are woven. Tomb magnificent with jewels and golden cups."

This was the most lavish tomb uncovered by Woolley, untouched by looters: that of "Queen" Pu-Abi, whose 4,500-year-old body was found buried between those of two workers; presumably the men had been poisoned before being interred with their mistress to continue to serve her after death. Among the well-preserved items of gold, silver, and lapis lazuli lay her intricately carved cylindrical seal, and this served to confirm Pu-Abi's identity.

By now, almost against his will, Woolley's work was gaining worldwide fame. This discovery attracted the same kind of public interest as that of Tutankhamen's tomb (which was a thousand years younger), also uncovered in 1922—it even inspired Agatha Christie's *Murder in Mesopotamia*.

The detailed cylindrical seal gave Woolley cause to believe that he had found an illustrious, indeed royal, character. But another seal unearthed by Woolley at Ur, less stirring in its simple, coinlike form with a few engraved characters, was misunderstood. Woolley simply labelled it variously as "Aramaic" or "Pehlevi." Later comparisons with other ancient texts have given the characters a fresh, better-informed interpretation as "Chemosh-nathan," the name of its owner. The first part of this name is the name of the god worshipped; the second part is the verb meaning "he has given." So its owner had worshipped Chemosh and lived in Ur in the seventh century BCE. Just as Yahweh was the god of the Hebrews, Chemosh is described in the Old Testament as the god of the Moabites who inhabited Ur. Thus the archaeological record was in accordance with the biblical text and provided important evidence to substantiate it.

So, Woolley and Lawrence had uncovered the original sites of Carchemish and Kadesh-Barnea, cities mentioned in the Old Testament that played significant roles in biblical history. Then Woolley had shed light on Ur and its Sumerian inhabitants, people who later became absorbed into a larger Mesopotamian culture as depicted in the Old Testament. This is all strong evidence that the historical tales told in the Bible are true, or at least based on historical fact. We have also seen evidence for Nebo-Sarsechim's existence around the time he was said to have been present at the siege of Jerusalem. The places where the people lived in these tales really did exist, along with at least some of the people themselves.

Of course, there are many more cases of archaeological evidence to support the biblical text. Within modern-day Iraq alone there are tens of thousands of archaeological sites with links to places mentioned in the Old Testament. In addition to Ur and the remains of its ziggurat (temple tower), Nineveh (located at Mosul)—a city founded by the biblical figure Nimrod and enclosed by 15 kilometers (about 9.25 miles) of walls—was revealed through the discovery of the library of Ashurbanipal. Here, around 1,500 different texts were uncovered, including the famous Epic of Gilgamesh with its flood story similar to that of Genesis.

At around 15 kilometers (about 9.25 miles) south of Kirkuk, the oil capital of northern Iraq, the biblical town of Nuzi—a center of Hurrian culture (which included the Horites, Hivites, and Jebusites of the Old Testament)—was located following finds of about 3,500 tablets dating from 1600 to 1400 BCE. These tablets describe ancient laws and customs, and help us to reconstruct social, economic, and legal practices in place just prior to the Israelite risings. They make reference to many Old Testament characters and places. Then of course there is Babylon, with its numerous references in the Bible, including five of its kings.

Babylon, also founded by Nimrod, was considered a biblical myth until in the nineteenth century its traces were unearthed some 90 kilometers (about 55 miles) south of modern-day Baghdad. These included the foundations of the Tower of Babel and of the royal palace built by Nebuchadnezzar, the Babylonian Chronicle, recording the capture of Jerusalem by Nebuchadnezzar in 597 BCE, and the Cyrus Cylinder, describing the fall of the city in 539 BCE. As Babylon—one of the most famous cities of the ancient world—emerged into the light, archaeologists viewed in awe the remains of ancient structures described in biblical stories. These stories were now no longer just tales but actual events, historical facts.

Having seen now the weight of archaeological evidence for the historical truth of the Bible, we are moved to ask: Why should the very first page of the Bible, the showpiece of any book, go against the grain? Indeed, does it in fact go against the grain? The truth we have uncovered elsewhere in the Bible is reassuring and encourages a belief in God: If part of it is known to be true, in that the people and places really existed, then maybe all of it is true, including the divine acts recounted. This further suggests that the original scribes were intent on recording the truth—the truth *throughout*, that is. It would have been quite out of character for them to have written something that bore no relation to the truth at the very beginning. But that leaves us with the thought that the creation story of Genesis surely cannot be the nonsense that the literal interpretation of its words suggests . . . can it?

This is where the science comes in. Can a comparison of the Genesis story with modern science provide more evidence for truth in the Bible? Let's take a look at that first page, beginning with God's famous first utterance: "Let there be light."

Sun

"And God said, Let there be light: and there was light."

God's first command in Genesis is "Let there be light." This is not the only introduction of light in the Genesis creation account, but as the first it must represent the formation of our solar system. This chapter will reveal the history of the science of astronomy, and how far this lay beyond the Israelites of old.

As the ancient Israelites observed the searing sun rise and set beyond the horizon of ragged hills or desert, they conceived of it journeying above and below the earth, which at the time was considered flat. The sun and the moon surely orbited the vast disk of land beneath their feet, they thought. Certainly, it is known that six thousand years ago the Sumerians believed that the earth lay at the center of the universe, and this concept was taken into Babylonia. Then, the ancient Israelites considered, the stars lay fixed on a heavenly sphere that enclosed all that existed. As we will discover in this chapter, the stars have many a story to tell; but so far as the ancient Israelites were concerned, the stars were nothing more than things of beauty and wonder.

GREECE TO GALILEO

Meanwhile, across the Mediterranean in ancient Greece, other ideas were developing, one of which was that the earth was, in fact, round. This notion emerged some time after the time the Pentateuch was written, when in the late sixth century BCE Pythagoras declared in writing that the earth, along with the other planets, was "officially" spherical. Pythagoras saw harmony in the universe, and the most harmonious geometric solid form is a sphere. It was noticed, too, that the shadow of the earth on the moon during a lunar eclipse was round. So, in Greek philosophical circles the earth was no longer flat. However, it remained at the center of the universe—and, as a consequence, so did man.

Having left Athens in 287 BCE, Strato of Lampsacus became head of Aristotle's Lyceum in Alexandria. A few years later, a parchment landed on his writing plinth—a thesis submitted by one of his students, Aristarchus of Samos. The crux of the thesis was expressed in a figure containing two concentric circles. On the first circle lay a planet—a spherical planet, in accordance with Pythagoras—and on the outer circle sat the stars. But there was something different about this figure, something that was in later centuries to be judged heretical, but that at this time would prove just too much to digest all at once. The spherical body lying at the very center of the concentric rings was labeled "sun."

Aristarchus of Samos' "sun-centered theory" proved unpopular. He was indicted on the charge of "impiety for putting in motion the Hearth of the universe," that hearth being the earth. Further, he had contradicted the affirmation of an earth-centered universe made several decades earlier by no less a figure than the great Aristotle—a risky business by itself. Eventually, around 150 CE, along came the Egyptian astronomer and mathematician Ptolemy (in Latin,

Claudius Ptolemaeus) to set the record straight. Ptolemy wrote a book called the *Almagest* in which he presented a great pile of data to prove the earth-centered model of the universe correct. Man's paramount place in the universe was restored, and everyone was happy. For the record, I should also mention Ptolemy's triumph in laying the foundations for making maps of various kinds; as for Aristotle, his greatness will become evident in the final chapter. But thanks in no small part to these two men, the earth remained at the center of the theoretical universe for another millennium and a half.

It was thought in the time of Ptolemy that if the earth did indeed move, then the air that we breathe would blow away. So, according to the deductive argument, since we were still breathing, then the earth was not moving. Ptolemy's planets, although different distances from the earth, existed on the same plane and so their orbits formed invisible rings like those that surround Saturn. Saturn, incidentally, appeared as the farthest planet from the earth, followed by Jupiter, Mars, the sun, Venus, and Mercury, with the moon closest. Uranus and Neptune were unknown before the introduction of the telescope in the early seventeenth century (discovered in 1781 and 1846, respectively). In the Ptolemaic system, the seven "planets" were termed the "wandering stars," and were positioned between the earth and the sphere of "fixed stars" that remained, which moved around the earth but retained their positions relative to each other. The fixed stars marked the boundary of the universe. The twelve constellations were picked out in the fixed stars, and were called the houses of the zodiac.

Although Aristotle and Ptolemy subscribed to a pagan religion, the Catholic Church found their astronomy pleasing. Further still, the fact that people from outside the Abrahamic religious framework had deduced a model of the universe that conformed to God's creation, in that it fitted with the biblical references, was taken as independent evidence for God's existence. Then a fly landed in the ointment.

In October 1632 Pope Urban VIII was perhaps busy in his study alongside the works of Ptolemy and Aristotle when a leather-bound tome thumped onto his desk. This was a case of supposition and retort as experienced by Aristarchus of Samos—but this time the outcome would be different.

The story begins with a brilliant German scholar named Regiomontanus. In the early 1460s Regiomontanus wrote a book called *Epitome of the Almagest* that popularized Ptolemy's cosmological view and added new observations to support it. Though it was not formally published for nearly twenty years, the book made Regiomontanus famous, and popular with the church. He produced horoscopes and made some sense of the houses of the zodiac—but also, more scientifically, he was able to make predictions of the phases of the moon and positions of the planets in the sky. This work was even used by Christopher Columbus to scare the hostile natives on Jamaica with his possession of "magic." With the aid of Regiomontanus' book, he predicted when a lunar eclipse would occur. When it happened as predicted, the impressionable natives became fearful—clearly, Columbus was some form of sorcerer.

In 1475, just a few years before he commissioned Botticelli and other Florentine artists to decorate the walls of the Sistine Chapel, Pope Sixtus IV summoned Regiomontanus to Rome. At this time astronomy was most useful for making accurate calendars, and Sixtus needed to know the correct Sunday for Easter and other holy days, since the current designations were more than a little dubious. Regiomontanus proved most useful, although to his own detriment: He met an untimely death as the plague swept through Rome. Still, religion and the subject of astronomy coexisted agreeably. But by now the idea of the earth-centered universe had become part of the new science of astronomy—a discipline that would progress relentlessly at, in its own terms, a logarithmic rate. In short, the model in which the planets revolved around the earth was about to receive scientific scrutiny.

Epitome of the Almagest was one of the first books owned by Nicolaus Copernicus, a couple of generations after Regiomontanus. Copernicus was born to a wealthy family in Poland in 1473, and was to become intrigued by the idea of the earth-centered (or geocentric) universe. He pondered it as he observed the sky at night, and decided to pit the theory against real data, which he collected as angles subtended by planets in the manner of Regiomontanus. He was particularly intrigued by a comment in *Epitome of the Almagest* to the effect that there must be some problem hidden within Ptolemy's theory of the motion of the moon. Exactly *what*, he asked himself, was this problem?

Copernicus studied mathematics, astronomy, and the obligatory astrology at Cracow University. He was a genius in the Michelangelo mold, a tormented loner who cared little for what others thought of him or his work. Quietly and remotely he made celestial observations from a turret situated on the protective wall of the cathedral of Frauenburg (now Frombork, Poland) where he was a canon, constructing precisely angled triangles from the positions of different planets on different nights. As he amassed his data, the information began to self-assemble into a picture of the solar system . . . *with the sun at its center*. From the angles subtended by each planet, it was clear that they could move in a perfect circle, and at a uniform speed, only if the sun lay at the center of the system. This was the same theory put forward by Aristarchus of Samos all those centuries ago; but this time it was backed by copious data. Now it really meant something—it could not be ignored.

Typically, Copernicus was not much interested in publishing his findings and attracting the world's attention. It took Georg Rheticus, a twenty-five-year-old German Protestant astronomer, spending two years at Copernicus' house and holding his hand through the writing process to get a manuscript ready for the printer. The world was about to be shocked. Man was about to be removed from pride of place in the center of God's universe. Copernicus put forward unquestionable evidence that the earth rotated once daily on its axis and traveled around the sun once yearly.

Nearing the end of his life in 1543 after suffering a stroke, Copernicus lay in bed in his substantial house for many days, his memory and mental alertness fading. But having finally been coaxed into preparing his work for public consumption, he hung on to see his magnum opus published. *On the Revolutions of the Heavenly Spheres* left the printer on May 24; just hours after taking a copy in his hands, Copernicus passed away.

The shock to the world took time to take effect, because at first few bothered to read Copernicus' work. Officially, the earth and mankind were still at the center of the universe. But among the small network of inquiring intellectuals, there were those who took note. One slight blemish on Copernicus' theory was a minor mismatch between the angular measurements actually made of the planets' movements and those predicted by his mathematical calculations. In 1605 another German prodigy named Johannes Kepler entered the story. Kepler understood that the original measurements taken were accurate, and that the discrepancy, of just eight degrees, could not be blamed on errors in conducting the experiments. He went on to propose the correct solution—that the planets' orbits were elliptical, as has been confirmed by later observations.

Throughout his astronomical endeavors Copernicus had striven to stay on the right side of the church, within which he lived and worked. He had no problem in reconciling his sun-centered (heliocentric) theory with the Old Testament scriptures. He saw God's magnitude and munificence in the exceptional beauty and harmony of the solar system—the principle of harmony, so important to the great Pythagoras, had been upheld. But this led him to a conviction that the planets' orbits must be perfectly circular, again in deference to Pythagoras. God, Copernicus reasoned, would construct only a perfect shape. The tone of his arguments had inclined the church leaders favorably toward Copernicus (although they did not go so far as to agree with him). He sought to get closer to God through revealing further details of His creation. But Kepler had broken that mold. Kepler's book, *Epitome of Copernican Astronomy* (1618–1621), a version of *On the Revolutions of the Heavenly Spheres* updated to include ellipses, became a scientific treasure, later to be used by Isaac Newton. But it also held information that would clash with Abrahamic religion.

Kepler was deeply religious himself, and was careful to circulate his book within the astronomical community only. After all, it was not so easy for the average person to digest, since it contradicted what they clearly observed—the sun moving around the earth from east to west. Further, although Kepler had brought the work of Copernicus to a wider audience of astronomers, few adopted his own idea of the elliptical orbit. A contemporary of Kepler, on the other hand, did away with all pleasantries and politics, and bravely took the whole sun-centered planetary system, complete with ellipses, to the world in general. This person was Galileo Galilei. It was *his* book that landed on the desk of Pope Urban VIII in October 1632. Astronomy was about to join astrology in the mainstream study of the stars.

Galileo was born in Pisa in 1564, also to a wealthy family. At the age of twenty-eight, after studying medicine, he became a professor of mathematics—his true passion—at Padua University in the

Republic of Venice. Galileo was a pragmatic Christian, less devout than Kepler. He believed that the Bible addressed moral issues, and that these were separate from questions relating to science and the universe that God had made. Yet he would fail, to his detriment, to convince the church that the two realms could be kept apart.

In 1598 Galileo revealed in a personal letter to Kepler that he was a believer in Copernicus' sun-centered solar system. But he kept this quiet for many years. Then, in May 1609, Galileo received a letter from a mathematician friend, Paolo Sarpi, reporting of a new discovery—a "spyglass." The spyglass made "visible objects, though very distant from the eye of the observer, . . . distinctly seen as nearby." The spyglass, of course, was a telescope.

I find it incredible that no one had invented a telescope previously. The Venetians had been playing around with glass for centuries, yet their skills had never been employed in astronomy. Greek philosophers had incessantly observed the stars with the naked eye, yet they failed to notice that the world viewed through a raindrop on a leaf was magnified. I'm sure I would have thought of it. Yes, *I* find this incredible; but that's because I cannot ever properly take myself back to the days of the Greek philosophers and Galileo, and to their mind-sets. Yes, I can imagine their daily life—their houses, clothes, and food—but I can never empty my brain of modern science.

This reflection helps to put into perspective the level of scientific understanding of the ancient Israelites who produced the creation account of Genesis 1. To put it bluntly, their sum of scientific knowledge, as expressed as a proportion of what we know today, was zero. The story covered in this chapter so far demonstrates that with zero scientific knowledge, it is difficult to make substantial scientific progress. Indeed, through the centuries, science has progressed stepwise, not in one fell swoop. Isaac Newton, who is itching to enter this brief biography of astronomy, commented: "If I have seen further it is by standing on the shoulders of giants." Anyway, at least we have recognized how those ancient Israelites were thinking when they produced their creation account—they hadn't a (scientific) clue.

Galileo used his skills as a mathematician and a workman to craft his own telescopes, superior to the original "spyglass." With these new tools he was able to see mountains on the moon, an observation that conflicted with the Aristotelian and biblical view of perfect "planets"—the view that Michelangelo had recorded in the Sistine Chapel. Galileo was also able to prove that the Milky Way was in fact made up of tiny stars, and found four small bodies orbiting Jupiter. In 1610 he published his observations in a book called *Sidereus Nuncius* (*Message from the Stars*, or *Starry Messenger)*. The book caused a sensation. And, since it was dangerously close to contradicting biblical assertion, it attracted the attention of the Inquisition.

The Holy Church had a quiet word with Galileo, advising that he distance himself from Copernicanism, although nothing official was broadcast. Clearly, the caution failed to have its desired effect—for in 1632 Galileo published his *Dialogue Concerning the Two Greatest World Systems*, presenting Pope Urban VIII with both Copernicus' sun-centered solar system and the first significant threat to the Christian Bible.

The *Dialogue Concerning the Two Greatest World Systems* is structured as a series of discussions, over a period of four days, among two philosophers and a layman. The layman is neutral, while one philosopher argues for a sun-centered universe—presenting Galileo's views directly—and the other for an earth-centered universe. The layman is left to decide between Ptolemy's view on astronomy and that of Copernicus. The intelligence of the advocate of Ptolemy was questionable—even his name, "Simplicius," was suggestive. The book was a classic case of a straw-man argument, and Copernicus came out on top. This was an ingenious way to get a point across to a mass audience more effectively than by dry science alone, although strictly speaking it would have been more scientific to have staged the discussion with two philosophers of equal intelligence. But it did appeal to the masses, and Pope Urban VIII, his own views represented by one "Simplicius," was forced to respond.

Galileo, who was not in the best of health, was summoned to Rome, accused of heresy, and eventually condemned to house arrest, for life, at his villa at Arcetri. He was also required to take an oath before the Holy Roman Church in which he renounced his belief that the sun lay at the center of the solar system.

Ironically, the church had been pleased with the work of pagans and outraged by the handiwork of devout Christians, working with church finance and support. These clashes marked the beginning of a split between the Abrahamic religion and science, which was to develop into a chasm over the coming centuries as science found its footing. The church had shown itself content to accept false logic if it provided evidence for God, a stance that persists in some religious circles today with the espousal of "intelligent design," as we shall see in the final chapter.

The rift that formed between science and religion during the debate over the position of the earth in the universe began a revolution in the notion of belief itself, and elevated thinking about the Old Testament to a new level. It further became apparent that what one believes, whether religion or not, demands the entire *energy* of the mind. Note that the concept of "energy" will become crucial in this and the following chapters. It will become evident in the electrons traveling within the atoms in the ink on this page, and in the light that strikes and reflects from those atoms so that we see them collectively. Energy is there in the sun, whence those light waves emanate, and in every star and planet in the universe, in which our sun is a mere grain in a salt cellar. Energy will emerge as a basic building block for . . . everything. Energy is special. And because we are concerned with how everything in our universe came to be, we must have an interest in where all that energy in everything came from in the first place. In the subsequent chapters we will follow the recycling of energy via the bodies of successive animals, and in the final chapter we will think of the energy that exists on earth as a limited resource that can indeed be recycled, but that must once have emerged from somewhere.

Energy certainly emerged as a common currency within science with the person who grasped Galileo's telescope—symbolic as the scientific baton at this point—with both hands. That person was Isaac Newton.

Newton, Einstein, and the Discovery of Gravity

Newton, an English mathematician, is fairly considered one of the greatest scientists who ever lived. It's hard to find any other single individual who laid more and better foundations for science than he did. Newton's discoveries are particularly important in our history of the universe as he answered the question *why* the sun was at the center of our solar system.

Born in 1642, Newton was not a particularly appealing character. Grumpy, irritable, malicious . . . one can continue along these lines. But such characteristics also indicate that his brain worked in unusual ways—he was not of the norm. Newton was one of the inventors of the branch of mathematics called calculus. He also solved the mysteries of light and optics, formulated three laws of motion, and derived from them the law of universal gravitation. His work underlies much of science even today—for instance, his laws of motion, put forward in his book *Principia Mathematica* in 1687, are the most fundamental laws of classical mechanics. These laws can be used to understand all interactions of force, matter, and motion, except for those involving "relativistic" and "quantum" effects, which were twentieth-century realizations.

In the late seventeenth century a number of theories of astronomy began to incorporate attractive forces, such as that of Giovanni Alfonso Borelli, and the concept of inertia (the power of resisting), such as that of René Descartes. This advance culminated in Newton's force-based theory of universal gravitation.

Newton was not, as the myth has it, struck on the head by an apple, but an apple may have been involved in his inspiration somewhere. The story may have unfolded something like this.

Upon seeing an apple fall from a tree, Newton began to think of the apple as accelerating, since its velocity had increased from zero. This led to his second law—that there must be a force that acts on this apple to cause this acceleration. The force is gravity, and the acceleration is due to gravity. Further, he must have thought that if the tree were twice as high, then the apple would still accelerate toward the ground. So the force known as gravity reaches even to the top of the tallest apple tree. But, he contemplated, how much higher can gravity reach? Could it reach all the way to the orbit of the moon?

If so, then the orbit of the moon around the earth could be a consequence of the gravitational force. The acceleration due to gravity could change the path of the moon in such a way that it followed an orbit around the earth. Imagine a cannonball fired horizontally from a mountain. The projectile will eventually fall to earth because of the gravitational force directed toward the center of the earth and the associated acceleration. Now imagine pouring more gunpowder into the cannon so that the cannonball travels much faster, although still fired in the same direction, horizontally. The cannonball would still be falling in the gravitational field, but now (possibly) at the same rate at which the earth's surface curves away. In other words, the cannonball never reaches the ground—it would have been put in orbit around the earth. Were it not for gravity, the cannonball would travel away from the earth and into space on a straight trajectory. Newton concluded that the moon was like the faster cannonball—it "fell" into its path around the earth because of the acceleration due to gravity.

Newton further reasoned that any two objects in the universe exert gravitational attraction on each other. The force has a universal form, symbolized as "G"—the universal gravitational constant. G is the same at all places at all times. But its effect is diluted by increasing distance and decreasing size, so that the closer one planet is to another, or the larger it is, the greater the gravitational force it exerts. Gravity will re-emerge later in this chapter as a force integral in shaping the solar system.

Unfortunately, one consequence of Newton's gravitational theory is that the stars in the universe are attracting each other and so must eventually collapse onto each other and draw in the entire universe. Newton was unable to explain why this did not happen. His problem was that he considered the universe to be static.

Although Newton's law of universal gravitation has since been superseded by Einstein's general theory of relativity, it continues to be used as an excellent approximation of the effects of gravity—indeed, relativity is employed only where extreme accuracy is required, or when dealing with gravitation for very massive objects such as stars. A solution to Newton's problem as to why the universe did not collapse in on itself followed. Newton had found that mass (m) and energy (E) were interchangeable; Einstein derived the first mathematical equation for the relationship, his famous $E = mc^2$ (where c is the speed of light in a vacuum).

Albert Einstein was born in Ulm, Germany, in 1879. After graduating from the Swiss Federal Polytechnic School in Zürich as a trained teacher in physics and mathematics, he found himself unemployed in his field. So he accepted a post as a technical assistant in the Swiss patent office.

We have moved from the ancient Greeks and their marble temples on the shores of the Mediterranean, through Renaissance Europe with its artistic genius and streets running with sewage, up to the modern age of patent offices. The scientists have passed through a wardrobe changing from full-length robes to frilled shirts, knee-breeches, and curly wigs, and on to shirts, ties, and suits. Since the Greek philosophers, inventions have included paper, eyeglasses, the abacus, the watch, and the printing press. The years following Copernicus saw the advent of the pencil, vaccinations, the steam engine, the camera, the bicycle, the telephone, and recorded sound. In Einstein's era came the lightbulb, the typewriter, the ballpoint pen, the radio, the car, the airplane, plastic, the television, and the electric guitar. Not that Einstein would have used *all* of these, although he did have an ear for music. That Pythagoras observed eclipses of the moon in a tunic and sandals, that Newton was wearing breeches when he observed his apple falling, and that Einstein sat as his desk pretty much as we do today, less a computer, enhances our glimpse into both the mind-sets of these individuals and the human scale of scientific progress.

To return to Einstein: While working at the patent office he began to formulate novel ideas in physics, including "relativity." This attracted the attention of universities in central Europe, between which he took to journeying like an academic gypsy until 1914 when he settled as a professor in Berlin. Two years later, Einstein finally proposed his general theory of relativity as a new theory of gravity.

Without going into the details of this theory, which would take up many pages, it is possible to note one important consequence, namely, that gravity was no longer described in terms of a gravitational "field" but envisaged rather as a distortion of space and time. Einstein's general relativity is based on his earlier theory of space and time ("special relativity"), which proposed that distance and time are not absolute—so that, for example, the ticking rate of a clock, or the length of a car, depends on the motion of the observer of those objects. In short, general relativity implied a universe that is not static.

In 1919, at the age of forty, Einstein became an overnight celebrity, thanks to a solar eclipse. An experiment had confirmed that light rays from distant stars were deflected by the gravity of the sun in just the amount he had predicted. With the planet exhausted from the First World War, Einstein emerged as welcome relief in the form of a modest scientific genius—humankind's nobility. Fatefully, he would become embroiled in the Second World War, but that's another story.

The implication of Einstein's general theory of relativity was that the universe was not static but must be expanding or contracting. Subsequent generations of physicists established that it was in fact expanding, and that it had a beginning—the Big Bang, some 13,700 million years ago. In the 1960s and 1970s three British astrophysicists, Stephen Hawking, George Ellis, and Roger Penrose, helped to define the Big Bang further by expanding Einstein's theories using measurements of time and space.

So far we have followed the progress in science from the sun placed at the center of our solar system because of gravity, to the nature of gravity according to Einstein, to the notion that the universe began at a measurable moment in time. At this point we are equipped to launch into an extremely brief history of the universe; and, most pertinent for our argument, we can look back to its beginning—that first point of being. We now understand what happened, although there remain a few little problems to solve. The following summary is necessarily rather full of facts: To make best use of it, other than understanding how the universe took its current form, look out for two other essentials—that everything is made from energy, and that the earth was left with a palette of building blocks called atoms.

THE BEGINNING OF THE UNIVERSE

We consider the Big Bang as an event lasting just 10^{-35} seconds—0.00000000000000000000000000000000001 seconds—in which the universe began with a cataclysm that generated space and time, as well as all the matter and energy the universe would ever contain. Energy is the lowest common denominator for everything in the universe. *Everything* can theoretically be broken down to energy, but no further. And *all* that energy we have in the universe today had just appeared in the Big Bang. In this 10^{-35} seconds—so short a time span it cannot be comprehended—the universe was a dense, blistering fireball. Also within this fraction of a second, a peculiar form of energy (what it was, we don't know yet) pushed out the fabric of space.

At this point there were no atoms in the universe. Atoms are considered the building blocks of matter, although, as I have just mentioned, they can be broken down further to energy. But it is convenient to consider the atom as a building block, because atoms can be joined together in different ways to make different types of matter, and even life. Atoms themselves are built from three types of particle: protons, neutrons, and electrons. Protons and neutrons—themselves further divisible into quarks and other particles—form the nucleus of an atom, while the much lighter electrons exist in clouds around the nucleus. Between the nucleus and the electron clouds is nothing. Atoms are more than 99 percent empty space.

The simplest atom is that of hydrogen, which has just one proton in its nucleus. Hydrogen is one of the elements, in that it represents one way in which the atomic nucleus can be configured. Today we have ninety-two naturally occurring elements, as represented in the famous periodic table. Each element has an atom with a specific number of protons and neutrons in its nucleus. But another stage of activity was needed before the universe came to contain atoms.

After the initial inflation, which lasted just one-millionth of a second, immediately following the Big Bang, the universe continued its expansion, but at a much reduced rate. As it expanded, it became less dense and cooled down. The basic forces of nature emerged—first gravity, then the strong force that holds the nuclei of atoms together, followed by weaker and "electromagnetic" forces. By the end of the first second in its history, the universe consisted only of energy and fundamental particles—quarks and electrons, and other particles such as photons (responsible for light energy), neutrinos, and less familiar types. Some of these particles smashed together to form protons and neutrons.

Then, in the following two seconds, protons and neutrons themselves came together to form the nuclei of simple elements—hydrogen, helium, and lithium. But these nuclei would have to wait another three hundred thousand years before they captured electrons into orbits around them to form stable atoms.

For the first ten thousand years in the history of the universe, most of the energy was in the form of radiation rather than matter. Radiation is energy that exists as "electromagnetic" waves or moving subatomic particles, and includes light waves, X-rays, and radio waves. This is the same "electromagnetic" spectrum that exists today, in which waves of shorter wavelength contain more energy that those of longer wavelength. As the universe expanded during its first ten thousand years, these waves were stretched and diluted within the increasing space.

At three hundred thousand years after the Big Bang, as the universe expanded further, atoms finally formed as free electrons linked up with hydrogen and helium nuclei. More energy now took the form of matter rather than radiation. We know this from the microwave background radiation, which hails from this time and is detected on earth today. While traditional optical telescopes view the space between stars and galaxies as pitch black, radio telescopes detect a faint background glow that is not associated with any star or galaxy. This primeval glow is composed mainly of microwave radiation—that stretched and diluted radiation with which the previous paragraph ended—which is why it evades our optical (light radiation) telescopes.

About three hundred million years after the Big Bang, even as the universe continued its rapid expansion, those regions in the primordial gas that were most dense became yet more dense, by virtue of gravity. Eventually, stars ignited within these pockets of dense gas, and groups of stars appeared to become the earliest galaxies. This information is evident in the pictures taken by the Hubble Space Telescope—a telescope lifted beyond the earth's "noisy" atmosphere by the space shuttle *Discovery* in 1990. At this point in the history of the universe, we are still something over 12,000 million years from the present.

Some 5,000 million years before the present, a giant cloud of hydrogen gas in the spiral arm of the Milky Way galaxy was rotating slowly when gravity acted to concentrate it all into one place. It started to rotate faster around its axis until a flat disk was formed, with most of the material in the center. The center of the disk became so hot and dense that it began generating energy through nuclear fusion of the hydrogen atoms to form heavier nuclei. It became the sun. After the sun had become a star in the universe, the rest of the material from the edges of the disk separated out into clumps and formed the planets, including the earth. The earth and the sun were born. *And there was light*.

I mentioned that two hydrogen atoms can fuse to form a single, heavier nucleus. This nucleus belongs to the second simplest atom or element in the universe—helium. So helium has two protons in its nucleus. Because these are both positively charged, and would tend to push each other apart, as equal charges repel each other whereas opposites attract, the helium atom also has two neutrons to make its nucleus stable. Such reactions take place within the searing cores of stars, including the sun. And the fusion process rarely stops here.

It follows that if three helium atoms are fused, the nucleus of the new atom will have six protons and six neutrons: This is carbon. If a fourth helium atom is combined with carbon, oxygen is formed. In this manner, all ninety-two elements are formed, often during different phases of a star's life.

Atoms of different elements join to form molecules, of which there are near-endless possible combinations. It is molecules that make up living tissue.

Phew!

So: It is out of light, in this case the boiling ferment of our embryonic sun, that all molecules are produced—including, importantly, those molecules necessary for life. Without light, in its most basic form, life would be impossible.

So we now also have the idea that the universe suddenly appeared at a beginning, and that it all appeared from a heck of a lot of energy. We are in the dark over what existed before the Big Bang, so we really don't know where all that energy came from. With the sun in place, nonetheless, we will follow the route that energy takes as it travels through time, taking on its various guises—mountains and oceans, air and animals.

Certainly, the ancient Israelites behind the account of creation in Genesis 1 would have been oblivious to everything described above, and indeed they provide no details of our solar system's formation. Probably I would begin a creation account with the formation of the sun and the planets. In the following stages of the Genesis author's narrative, on the other hand, an order of events begins to emerge that is more surprising—because it is surprisingly accurate, scientifically. The specific order of events, and the phrases used to describe them, bear the hallmarks of modern scientific truth. The two perspectives on the history of the universe, and later of life—the scientific and nonscientific—although widely divergent, are beginning to converge on the same story line.

The true building blocks for life are now in place. We have seen how some immeasurable amount of energy became transformed into the entire universe of stars and planets that we see today. At the point we have just reached, the sun is in place at the center of the solar system, but the planets, including the earth, have only just come into being as clumps of material that once existed on the edge of the sun in its early form.

To put the creation of the sun at the beginning of the story of life is perhaps not surprising; it makes intuitive sense. So the fact that Genesis and our modern scientific narrative agree on this point might not seem that striking; but, as we will see as our story unfolds, the agreement by chance between science and the Bible on the next steps is increasingly unlikely.

Gravity pulled the matter of the earth together to form a sphere, but a sphere unrecognizable as that of today. Further physical processes set about molding and changing the young planet, and these will be the theme of the next chapter. From here, we will follow the voyage of energy on earth.

SEAS

"And God said, 'Let the waters under the heaven be gathered together unto one place, and let the dry land appear.'"

According to the first page of Genesis, God created a "firmament" that he called heaven, and divided the area below heaven—the earth—into land and sea. The formation of land and sea ties in with our scientific understanding of the first stage in the earth's history.

In chapter 1, I mentioned that Woolley and Lawrence noted "bands of flinty strata" on their journey in North Sinai. By their time, in the early twentieth century, the term "strata" meant something scientific. It was an indication that the hills they traversed had a *very* ancient origin. Although unfamiliar to the Israelites of old as they trod the very same rocks, that origin, and how *we* came to know it, will be a theme of this chapter. This is all about the earth's very early history, when some of that energy generated in the Big Bang manifested in minerals and seas—when hydrogen and oxygen atoms combined to form water.

To understand how the seas were formed we will need to look

back into the depths of time, to examine the fossil record and see what this tells us about the earliest conditions on earth. We must then consider environmental forces such as volcanic action and weather, and how these may have set about shaping our embryonic planet. But to begin we must look at another moment in this history of science—when someone for the first time postulated a sequence for the formation of the earth.

The Earth Gets Older

While Galileo was being condemned by the Roman Catholic Church in Italy, René Descartes was fretting over his own fate as he contemplated publishing his scientific discourse on the shaping of the planet—a thesis equally challenging to the literal interpretation of the Bible. Descartes was born in France, but had moved to the Netherlands where he spent over twenty years doing work that would revolutionize mathematics and philosophy. He also made contributions to *natural* philosophy (what we call science today) that are relevant to this chapter. One of these was his proposal that the earth's formation had been a historical process.

Clearly this did not sit well with the church, since a direct translation of the Bible reveals a rapid, seven-day creation. But for the purpose of this book we must remain rational, and refuse to accept the seven-day timescale. Of course, coming into an intellectual environment based for centuries on the belief that the Bible's every word was true, Descartes' ideas were hard to take on board, certainly among his clerical contemporaries. Today we are no longer bound by such constraints and can afford to read the Bible more metaphorically. So what did Descartes say on geology, exactly?

Tactfully, he suggested that the present state of the earth was a result of natural laws governing the development of the original matter over a long time span. Revolutionary, spot-on, but brief—and, so far, without supporting evidence. It took Georges-Louis Leclerc, comte de Buffon, in the eighteenth century to provide the data Descartes' theory desperately needed, and to kick-start the science of geology.

Buffon was director of the royal botanical collection in Paris, and became an outstanding figure in the history of science. He became convinced that the planets had a very ancient origin, much beyond the six thousand years inferred from the Hebrew Bible in the previous century by Archbishop Ussher (who placed the first day of the creation at nightfall preceding Sunday, October 23, 4004, BCE, having deduced a date for Adam, created on day six). But Buffon's first three books had been eating away at the church's patience, and at the point when his fourth was ready for publication in 1753, he took a leaf from Galileo's book and dissociated himself from his own views. Buffon declared that he believed in the literal truth of the creation account in Genesis, and that his theories on how the planets formed—in which he was approaching the modern scientific view—had been only "a purely philosophical supposition." He had been merely experimenting with thought. But he would quietly continue his experiments.

In 1779 Buffon wrote: "It is not possible to doubt . . . that there have occurred an infinity of revolutions, of upheavals, of particular changes, and of alterations on the surface of the earth, . . . by the actions of rains, of frosts, of running waters, of winds, of subterranean fires, of earthquakes, of floods, etc." He was, of course, correct; but, as everyone knows, these environmental processes act slowly. Buffon had no option but to make an earth-shattering, and certainly religion-rocking, prediction—that the earth was seventy-five thousand years old, and indeed possibly much older still, even millions of years old. Such units of time were not in the vocabulary of the age; no one had had reason to conceive of a *million* years before.

Buffon's ideas were heavily influenced by fossils. A fossil is an individual animal, plant, or more primitive life form preserved in rock. Whenever a marine animal, for instance, met its demise, provided it was not eaten by others it would be buried on the seafloor by sediment falling out of the water above. If it became buried before bacteria took hold and decomposition fully set in, its body shape would be preserved within the sediment, which in turn became rock. As the rock passed through physical and chemical changes over millions of years, the animal, now a fossil, would take on a new mineral composition too. The upshot is that today we can break open rocks of different ages and find the likenesses of animals and plants that lived at the time the rocks were formed. The organisms' bodies have become immortalized. Step by step we build up an increasingly thorough fossil record, or a *column* of fossils from deep to shallow within the earth's surface—from old to young within geological time.

That's a view we take for granted today. But with no fossil column within which to work, Buffon had to consider how to treat fossils with the scientific method. What hypotheses could be made and tested against real data? He lined up fossils from different depths in the rock and realized that they formed groups of communities of animals; indeed, that each level contained its own specific community. He was in fact constructing the *first* fossil column—a time sequence in which fossils would have formed, as they might appear in a bore sample. Clearly, life was different at different periods of time. Buffon assigned each community group to a distinct "epoch" and identified seven different epochs in total. Life had passed through seven stages of change, presumably caused by changes in the planet itself, and this must have taken a long time—*geological* time. Buffon had found evidence to support Descartes' idea of an *old* earth.

Today we are blasé about such an idea—to say it is common

knowledge would be an understatement. But two and a half centuries ago, the one and only model rattling around in our skulls would have been that the earth and all its life forms were created within a single week, six thousand years ago. So Buffon's notion of an old earth would have been hard to absorb. Not even the identification of seven stages in life's history, as designated by Buffon, would have helped to ease the transition from biblical literalism to natural law—seven being a comforting number to religion, employed also by Isaac Newton when he set up his spectrum. To bridge the gulf between the old assumptions and the new science, a whole new attitude and a new politics were needed among the population. That's why Buffon was so important to the history of science. He laid geological foundations and paved the way for others.

Buffon and his contemporaries were on thin ice with the political and religious establishment of their day. But in the nineteenth century the psychological dam built by the church to stem the flow of geological discovery finally burst under the pressure of a torrent of independent evidence from myriad scientific studies. It became personally safe to engage in scientific study of the natural world, and the Victorians in Britain, and their neighbors on the European continent and in North America didn't hold back. Progress was rapid as the new model scientists made up for lost time—rocks and fossils would surely have been understood earlier, perhaps back in the Renaissance, had it not been for the blockages and disaproval of religion. But that's *religion*—the human interpretation of God, heavily influenced by certain personalities—not God Himself. It is perhaps ironic that Darwin would become the target of religious criticism when, had it not been for religion, natural selection would probably have been expounded before his time. Maybe Darwin would have discovered DNA? Anyway, Darwin must wait until the next chapter.

While some scientists condemned fossils as the work of Satan, planted with the intention to deceive, many others set sail from the port of natural laws, assuming the mantra that natural laws have always ruled *all* of nature. Between these two extremes nestled a third group, the rationalist "deists" particularly common during the Enlightenment, who held that, at the moment of creation, God created both matter and the laws of how life would work. Laws of nature. The deists deemed further that science would reveal these laws as evidence of God's activity. Here I must mention Darwin one more time, since he was for at least part of his life a deist.

Cycles and Catastrophes

The science versus religion debate in Britain at the beginning of the nineteenth century was conducted not between extremes, as it is so often today, but between moderate Anglicans and deists—both of whom were well represented in scientific circles. The moderate Anglicans believed that the new evidence of geology was actually compatible with an interpretation of the Bible that included a *literal* interpretation of the beginning of Genesis. In the words of the then president of the Royal Academy of Dublin, "geology naturally ripens . . . into religion." He believed that the scientific data on how the universe and life began would paint a replica picture of the seven-day event. Two British deists—James Hutton and Charles Lyell—provided evidence to challenge this position.

The Scottish scientist James Hutton lived through most of the eighteenth century and laid down a rule that would be crucial in understanding how the earth came to be as it is today—that the past events of the earth's history can be explained in terms of processes still at work today. This, his "principle of uniformitarianism," freed scientists from the need, or impulse, to factor into their theories miraculously powerful events that no longer occur—events that could not be conceived of, let alone quantified. Geology was no longer required to work around an accepted black hole in human knowledge. Newton had claimed that the physical laws of motion were the same everywhere in the universe. Hutton added that they also remained constant throughout time. Like Buffon, Hutton had revealed an error in the concept of the six-thousand-year-old earth. The forces acting today could only have shaped the planet through very gradual changes, and that would have taken *considerable* time. Just to begin, think about how much time it took the Colorado River to carve out the Grand Canyon from the rock.

Taking Buffon's idea of epochs and incorporating it into his own theory, Hutton formed a view of the earth's history as a series of endless mechanical cycles. Oceans retreated to leave land, and then the land sank to lie beneath the ocean, so that the land of most countries has spent time both above and below the water. Rocks weathered and were reduced to sediments that were buried in water, and then the sediments would be compacted to form rocks, which during geologic upheavals were lifted up into the air to become mountains. The mountains would erode and the cycle would continue. Hutton's theory of cycles contrasted with Buffon's theory of constant change, and was reassuring in that it meant people no longer needed to anticipate an unknown, possibly devastating force lurking around the corner.

Charles Lyell was born to a prosperous Scottish family in 1797, the year Hutton died. That was appropriate, since Lyell would carry Hutton's cause into the era when science got serious—the nineteenth century.

Lyell switched from law to become professor of geology at King's College London in the 1830s. This was an interesting period for British science, one in which the notion of natural selection was awakening within two minds either side of the Welsh border. Although I have agreed to save that topic for the next chapter, it's worth noting here that the books published by Lyell would have a direct influence on the evolutionary pioneers. His publications were important in that they brought a century and a half of ideas on the earth's development into the public domain.

Lyell's first book, *Principles of Geology*, published in three volumes between 1830 and 1833, propounded the doctrine of uniformitarianism. He had traveled widely across the world and had made numerous observations, for example from active volcanoes, that planted steel supports in Hutton's precariously balanced model. He also tweaked that model a little, pointing out that although discrete epochs did exist, nothing on the earth changed to drastically alter the *way* that life has always worked. Considered as a totality, life has remained always roughly the same. The fossil record, on the other hand, although thin on the ground at this stage, did reveal an uneven and progressive development. Lyell put down this apparent contradiction of his views to the same misrepresentation we would gain from studying a living population of a single species today, where fluctuations take place with each generation. Each generation, in this case, would offer a different snapshot of the population—and fossil assemblages are indeed snapshots of the past. It took the next stage in the history of science to correct Lyell's view and establish that the changes in the fossil record are not the same of an order of magnitude greater than those observed through the generations of a species.

Hutton and Lyell also shared the same view on religion: They believed in God, a deity, but not in divine intervention or the revelations ascribed to Moses and the other prophets in the Bible. Hutton, for instance, believed in a God who had created the matter in the earth and had set the physical processes in motion, but who did not intervene thereafter. The idea that Hutton left us with was that the earth's surface was subject to two basic processes: That rocks were worn away by weathering and erosion, and that they were then re-formed and uplifted by heat and pressure. Now people knew why marine fossils were found at the top of mountains. This, of course, clashed with the religious perspective, in which fossils found on mountains could be seen as evidence that the earth was once completely submerged by the great Flood (unfortunately some creationists still try to make this fallacious argument today): When the deluge ended and the waters retracted, the aquatic animals were left stranded and became fossils on the land. Then along came French scientist Georges Cuvier who, in the early nineteenth century, discovered that the rocks of the Paris basin revealed not just one but a *series* of catastrophes during the earth's history. Cuvier did not know what these catastrophes were, although he did date the earliest to six thousand years previously, so that it conformed to the formation of the earth in the scripture.

In 1821 Oxford University geologist William Buckland also used science to support the Bible. Buckland, himself a clergyman, discovered bones of extinct animals in the Kirkdale Cave in Yorkshire that had apparently died in a sudden flood. Buckland argued that had we not heard of Noah's Flood, we might have deduced it through science. He went on to invoke further catastrophes to explain other fossil assemblages, moving toward ice as an explanation rather than further floods. Interestingly, among his published record of fossils lay the first description of a dinosaur, although the term itself would not be coined until a few years later, by Richard Owen.

Owen was a British scientist best known for his work on the anatomy of animals, and detailed comparisons between them. From his meticulous knowledge of bird anatomy, he once identified the giant extinct bird called the moa from one tiny bone. Unfortunately, he also became synonymous with flawed criticism of Darwin. It was Owen who promoted the building of a "cathedral to natural history" in London, which became the Natural History Museum. Ironically, this institution has since become a shrine to natural selection, and Owen's marble statue has recently been ceremoniously replaced with that of Darwin, his own being demoted from the head of the staircase to the canteen below. But it should be remembered that, for all his misguided and on occasion vindictive hostility to some of his contemporaries, Owen laid admirable foundations for the field of comparative anatomy. Indeed, the art of scientifically describing and cataloguing animals is completely removed from the subject of evolution: Evolutionary study begins with this information to make its predictions of ancestry.

The scriptural scientists did not have it all their own way. One blow was struck by Scottish scientist John Fleming, who rained on the parade somewhat by pointing out that the scientific evidence supported only local flooding, not a universal deluge. He went further to show how the catastrophes they had uncovered, separated by *vast* periods of time, contradicted rather than confirmed Genesis. Although he set out to contradict Hutton too, as Fleming proceeded his thoughts tipped from Cuvier's catastrophism toward Hutton's uniformitarianism as an explanation of the planet's history. He also endorsed Buffon's view of a sequence of major epochs in the history of animals, although he pointed out that the correct number of divisions was in fact six, not seven (just as there are in fact only six colors in the Newtonian spectrum, indigo being a superfluity added to produce the "correct" number). How irreligious. No wonder his work was blocked.

But by now there was a mounting mass of evidence for an old earth. The next piece of the jigsaw fell into place in 1858, the year when the French geologist Antonio Snider-Pelligrini publicly declared that the earth's continents were once joined and have since gradually drifted apart, and when the curious, conspicuous detail that the continents could be fitted together like a jigsaw puzzle—most obvious when looking at Africa and South America—finally rolled off the printing press. This idea of continental drift developed into the scientific theory of plate tectonics. The fossil record soon added to the tectonic evidence, for example with the discovery of warm-climate plant fossils lying near the poles. Clearly, the land masses on the earth's surface had moved around through geological time. The separate plates, floating on a bed of molten rock, may crash into each other or move farther apart on their slower-than-a-snail's-pace travels. A plate that once lay on the equator and supported tropical life could today be found at one of the poles.

Seas were factored into the first models of continental drift, and appear a prerequisite in Hutton's theory of cycles. Seas cover any part of a plate that sinks low on the earth's crust, and so reveal the outline of rocks above a certain height around the globe—the outline of the continents. This in turn revealed the shape of the pieces in earth's jigsaw puzzle and pointed the way to continental drift.

The formation of the sea as well as the land is chosen as the second stage in the creation on the Bible's first page. Modern science reveals that land and sea certainly were in place before the next stage in the scientific account of the history of the universe—the appearance of life—as I will show. But so far we have confirmed only that the Bible's inclusion of the division of the earth into land and seas was an appropriate and worthy addition to its creation account. Now let's ask whether it is also in the correct place in the sequence of events.

We westerners came to understand the earth's formation while causing minimal damage to the scriptures. Through this process we arrived at a fairly accurate description of how the planet came to be as we know this prehistory today—through a series of catastrophic events alongside continual, more gradual change, all the while subject to physical laws that were never unlike those in place today. We know that land moved around both horizontally and vertically on the planet's surface. And we know that all this took place in *geological* time, which does sit awkwardly with the Bible.

We have also reached the year 1858, and that's a landmark date. It was the last year that science and religion would remain friends, or at least speak the same language. The reason for this will become clear in the following chapter. But before leaving this chapter, and before closing with an update on the earth's formation, it will be interesting to apply the geological ideas encountered so far to a particular case—and what case could be more appropriate than the land that supported the ancient Israelites?

Woolley and Lawrence observed strata at first hand in the North Sinai, for example at Nagb el Safa, where the layers exposed at the surface of the hillside had become broken and loose. Thin plates of slatelike stone moved under their feet, and made hiking tricky.

On their journey Woolley and Lawrence, like the ancient Israelites, would not have failed to notice the deep gouge in the earth's surface that extends from the southern end of Turkey into Africa, tearing the Bible lands in two. The gorge supports the River Jordan, within its cavernous valley, and several lakes, including the Sea of Galilee and the Dead Sea. In the 1920s a group of British geologists led by John William Evans—a trained lawyer who turned successfully to science—investigated the Judean desert. This expedition was undertaken by order of the Palestine Mining Syndicate, a group formed to administer and monitor the new search for oil in the Palestine region, while the necessary maps were provided by the Palestine Exploration Fund. Evans and his colleagues came to the following conclusions.

The rocks represented in the Judean desert belong to the Middle and Upper Cretaceous period, between 125 million and 65 million years ago (although we have reached nowhere near that far along in the history of the earth so far). That is, all that time ago, the rocks were formed by particles of sediment falling out of a body of water and onto the bottom. As the pressure built up on the sediment that had settled, it became pressed into a solid layer—that is, it became rock—as fluids were squeezed out. The chemistry of the layer of rock changed over time, so that today it may contain elements different from those present at the moment when it first formed. Also through geological time, the rocks changed their position on the globe. They became raised above the water during the gradual uplift of the area—as those tectonic plates moved around and forced some seafloors up into the air.

At the surface of the Judean desert three types of rocks are identifiable, known as "strata," or "beds." A stratum, or bed, is a thick layer of rock formed during a specific period of time during which environmental conditions remained relatively stable. The beds from different periods were successively piled on top of each other, so that the one that formed first—the oldest—is found at the bottom of the pile. An animal buried while a rock layer is forming will eventually appear as a fossil in that bed. It was the transitions between beds, taking place so suddenly (in terms of geological time), that led the early geologists to infer that on each occasion a catastrophe must have occurred to change the environmental conditions so severely and abruptly.

The whole Judean desert was under water for the entire Cretaceous period, although the type of water above it changed. When the lower beds were formed, they were below open sea. Then the water began to recede, and the upper beds were laid down in shallow water. By the time the uppermost bed was deposited, the water was a mere lagoon. Unlike the waters that formed the lower and intermediate beds, this lagoon was lifeless, which we know because it left no fossils.

Today, the *surface* of the entire Judean desert appears as a mosaic of these three beds. Consider the three-layered rock—made from the three beds—that, after it is lifted above the water, is worn away by weathering over time. Where it is little eroded, the top layer remains at the surface. Where it is most eroded, the top and middle layers are removed to leave the lower layer exposed. But where there is a clean break in the nonweathered rock through a geological fault—a landslip—the part that is pushed into the air reveals a cliff face with all three layers exposed in cross section. These are the strata referred to by Woolley and Lawrence, visible in the stripy appearance of the cliff faces or mountainsides. The slatelike rocks encountered by Woolley and Lawrence on their hillside were even thinner layers—sublayers from within a single bed—that were breaking away from the surface along their horizontal weak points.

The deep gouge that contains the River Jordan was once attributed to landslips. But John Evans' geological expedition came to a different conclusion—that this was a major fault line. The earth's crust in this region was being moved in opposing directions by activity beneath, and this was the place where something had to give: So the crust tore apart. This was an action of tension rather than compression—where the crust is compressed, for example, where the Indian plate crashed into the Asian plate, mountains are formed (in this example the Himalayas). The Judean fault line began some twenty million years ago, and its extension remains in progress today, which accounts for the numerous earthquakes still felt in Palestine. Land had been displaced by as much as a thousand meters (about 0.6 mile) on this line.

Today, the whole of the Bible land and southern Arabia has been mapped, and everywhere the story is similar. Again, the rocky desert today was once, in geologically ancient times, the bottom of a primordial sea. But in southern Arabia, the rocks exposed at the surface derive from even older periods in geological time. Interestingly, there is a region some 500 hundred kilometers (about 310 miles) south of the Dead Sea that derives from a period known as the Cambrian—an age of immense importance in the history of life, as will be explained in the following chapters.

So far I have mentioned two periods in earth's geological history, the Cretaceous and the Cambrian. The earth's history has been divided into many periods, and what is known as the Cambrian lasted from 543 million to 490 million years ago. Seas existed well before this period—before the land of the Judean and Arabian deserts had formed. So how did water come about on our planet?

We have reached a stage in our understanding where the earth is in place and beginning to take its familiar shape, arriving at this picture by evoking the natural processes found on earth today. We have followed the trail of scientific pioneers to explain how, given enough time, land can form from sediments settling out of the sea and being thrust upward into the air from below. While sidestepping accusations of heresy, those early scientists revealed cycles on the earth that worked endlessly to maintain long-term stability.

We have yet, however, to account for the liquid that even today covers over 70 percent of our planet's surface. This is the point where I need to bring the story of the earth's formation up to date with modern science. From here to the end of this chapter, I will describe the series of events that must have taken place during the earth's development according to our modern understanding of gravity, chemistry, and how materials react under different conditions. Since we know the end point—the earth today—and how the chemical elements on the earth are capable of changing, we can postulate exactly how they did change, and what conditions must have occurred to prompt those changes.

THE COOLING EARTH

The previous chapter ended with the earth in place in the solar system, spinning on its axis and orbiting around the sun. It had formed from a clump of material that was once part of the sun, and contained a variety of chemical elements. As our proto-planet developed, it was always in danger of being struck by one of the other planetary bodies that hurtled around in the same pocket of space. The inevitable happened, and the earth was hit by a body about the size of Mars. It was hit so hard that the debris blasted off the surface formed the moon.

The earth was born hot—molten hot. But the sun, at that time, generated 70 percent less heat than it does today, which allowed the earth to cool. An atmosphere began to form from rock vapor escaping the molten surface, although this vapor only crystallized and fell back into the seething surface. The cycle continued for a few thousand years until the earth began to lose heat, through convection, into space. Powerful convection currents churned the earth's body and the denser elements, such as iron, began to sink to the core. As the earth's heat became concentrated in the core, a cooler "mantle" formed on the surface of the planet, although the crust had yet to materialize.

At this point, the earth's surface and atmosphere would have been about 1,800 degrees Celsius (about 3,272°F), but the atmosphere began to change. It became dominated by carbon dioxide and . . . water vapor. Hydrogen and oxygen molecules broke their bonds, and their atoms combined to form water. This was indeed a fortunate event, not least because the earth's hydrogen might have been lost from its atmosphere had it not been locked into water.

Thanks to gravity, the earth held on to its early atmosphere. The surface began to cool to form a thin, hot crust. Within just a few hundred years the crust was sealed off from the material beneath, and the entire surface of the planet became solid, except for volcanic moments where molten lava from beneath forced a fissure, and where asteroid impacts broke it open in places. The crust was composed of material known as igneous rock, formed by the solidification of molten rock, or magma. This is different from the sedimentary rock mentioned above, which formed from sediment falling out of the sea, although both igneous and sedimentary rocks can change their chemical or physical characteristics, under high temperatures and pressures.

The dense atmosphere trapped some of the sun's radiation, although much more heat radiated out into space. The earth was losing more heat than it was gaining, and the net result was that the earth cooled to a mere 200 degrees Celsius (392°F). At such a temperature water vapor can liquefy, and it began to rain. The rain, which carried carbon dioxide, began to gather in low regions of the crust, and shallow, hot, acidic seas appeared.

Volcanic action and large asteroid impacts supplied energy to fuel chemical reactions between the elements. Sodium and chlorine condensed out of the atmosphere and combined to form salt in the early oceans.

A few hundred million years later, the crustal rocks sank into the mantle. The earth's surface cooled further and became increasingly dense, and sank deeper into the interior. The surface rocks carried away the carbon dioxide they had absorbed from the water into the molten rock below. The dense early atmosphere, along with its greenhouse effect, was being removed. Recycling of minerals from cold to hot regions was taking place, and a suite of new molecules resulted from the ever-changing conditions.

The energy that the earth had been handed at its embryonic stage had become unrecognizable. The earth remained a vast ball of energy, but it had been converted to many different forms. Hard rocks and water characterized the surface, with molten rock below and a unique gaseous atmosphere above. Lots of atoms, with their animated electrons and ample nothingness, were being joined together in a variety of ways (as mentioned in the previous chapter, atoms are more than 99 percent empty space). This would lead to just the right mix of carbon, hydrogen, nitrogen, oxygen, phosphorus, and sulphur for . . . life.

By 4,350 million years ago, some 250 million years after its formation, the earth possessed mountains and seas, with a water temperature of just 30 degrees Celsius (86°F) at the equator. The sea was salty and, like the rocks, it held minerals of various kinds. While Venus never cooled enough to have liquid water, and Mars and the moon were too small to hold on to their atmosphere long enough for its elements to be stored in their mantles, the earth sat pretty in the solar system. It was just the right distance from the sun to retain its water mainly in liquid form, rather than as ice or vapor. Quite simply, the earth got all the good stuff. A cradle for life had been prepared. Still, no life so far.

Before moving on enthusiastically in the following chapter to the next big step in the earth's history, I should mention an interesting coincidence involving the sun's output and the transparency of seawater. The sun has a peak in its energy output—in its electromagnetic spectrum—and seawater has a peak in its ability to transmit this electromagnetic spectrum. Strangely enough, and there is no scientific reason for this, the two peaks converge on the same narrow window of wavelengths—the visible light range. Considering the vast range of wavelengths in the sun's output, it is indeed a mighty coincidence that this minuscule region is also that which travels best through seawater. This will prove important, even fateful, in chapter 6.

In this chapter, again, we have seen how centuries of science accumulated a mass of knowledge sufficient to solve the mystery of the earth's formation down to the finest details.

So far, the author of Genesis has described two events in the formation of the earth. That he devoted one of his few stages in the creation to the sea is interesting, since the sea, as we will find in the following chapters, was necessary for life itself and the diversity of modern life. Yet the writer of the creation account of Genesis 1 lived in a landlocked region, and even Moses tended to opt for the desert over the coast. A similar situation existed within the Mayan civilization of Central America, whose people relied upon freshwater pools, but not surprisingly were oblivious to the science of these pools. Actually, they were all connected—part of an underground freshwater river that was exposed at the surface here and there to reveal the pools. The Maya, nonetheless, did not produce a text that described the underground water system. The ancient Israelites produced Genesis.

We have now seen the formation of the sun ("Let there be light") and the separation of land and sea, and both are found in their right sequence in science and in Genesis. The story may seem rather predictable so far; but with the next stage in the earth's history described by the writer of Genesis, as we shall see in the following chapter, the odds in favor of guesswork begin to fall. Finally, it's time for Darwin.

CHAPTER 4

LIFE BEGINS

"Let the earth bring forth grass, the herb yielding seed, and the fruit tree yielding fruit after his kind."

The opening page of Genesis asserts that plant life appeared after the seas were formed, and names specifically grass, herbs, and fruit trees. What is really implied here is that this is the stage in the creation process where *life began*. It is the first phrase in Genesis that mentions life of any kind. As we will discover in the following chapter, the simple forms of life that did first exist were not discovered until a couple of millennia after Genesis was completed. So why did the writer of Genesis pen, specifically, grass, herbs, and fruit trees at precisely this moment in his creation narrative? Is it appropriate for such plants to appear at this point in the creation account, after the land and seas had formed? Do these plants bear any relationship to the scientific understanding of the early history of life? This chapter will attempt to find answers for these questions.

This is the chapter in which we will consider how and when life began, and how it began to diversify—the origin of life and evolution. Indeed, science needed evolutionary theory—to divert attention from the creation—before the question of how life first appeared could be posed. To comprehend the actual, scientific history of early life on earth, it will be helpful to consider how we came to know about it—the various stages, including false starts and dead ends, in our understanding of the subject. This chapter will recall a story of a small red book and two green ones. We must start, however, with a tall but slim maroon tome—Erasmus Darwin's *The Temple of Nature*, published in 1803, the year after his death.

Erasmus Darwin was a British physician who developed into a polymath. He wrote extensively on medicine, zoology, botany, engineering, and philosophy, as well as composing poetry—he deeply influenced Blake, Wordsworth, and Shelley. He was one of the founding members of the Lunar Society of Birmingham, a discussion group of pioneering industrialists and natural philosophers. In the late eighteenth century he began to develop his ideas on evolution, which culminated in *The Temple of Nature*.

Today, when we talk about the "tree of life," the point is not so much about the specific branches as about the existence of closely related groups generally. Fishes form a group of their own, as do insects. Within the insect group are smaller groups—bees, beetles, and butterflies. Breaking down these groups into even smaller groups is a task of science today, and sometimes we get the groupings wrong before we get them right. But the point, again, is that the hierarchy of life is a universal principle of the natural world, and by far the most likely explanation for this is evolution. This is what Erasmus Darwin hinted at. He believed that life had developed from microscopic specks in primeval seas through fishes and amphibians to "humankind." He did not correctly explain the mechanism by which this proposed development occurred—he did not even come close to the theory of natural selection, which in many ways is the clever bit in evolutionary theory. But simply arriving at the concept of evolution, amid minds exclusively assured that God created all the species, was revolutionary.

Erasmus Darwin, writing in sensitive times with the French Revolution under way just across the Channel, would always have half an eye on the church and the nervous disposition of the establishment. He held back where others would later dive in, and avoided presenting society with anything too radical to digest. Indeed, he even cloaked his evolutionary inclinations by composing *The Temple of Nature* in the form of a literary work, a poem. This is one of the ways in which Erasmus Darwin threw a veil over himself, somewhat concealing his genius from the history books.

There were certainly people ready to pounce on Erasmus Darwin if ever he drew too much attention to the theory of evolution. Not least among these was William Paley, an influential Christian philosopher who published prolifically on natural theology (a branch of theology based on reason and ordinary experience, rather than on scripture and religious experience). Paley came up with what he asserted was evidence that the world was designed and sustained by God—mangling, in the process, the new facts emerging from the scientists' endeavor. His most famous argument involved a comparison between a watch and the world. This holds that the complexity of a watch could never have come about through random chance or through any natural process, so the only possibility is that an intelligent watchmaker must have designed it. By the same reckoning, the towering levels of complexity found in the universe or in an animal must also have been designed by an intelligent designer—God. (It is, of course, from this idea that we derive the phrase "intelligent design.") The principle was an old one, first put forward by the Roman statesman and philosopher Cicero. It is deeply flawed, not least because it is teleological—it presupposes that there is a purpose or directive principle in the works or processes of nature, rather than beginning with a blank slate devoid of belief or preconception. Evidence is not considered impartially.

The watchmaker argument falls down completely if there exists a plausible explanation of the universe or an animal's anatomy in terms of natural processes. Suddenly, God is not needed in the explanation. The teleological argument is fatally vulnerable to advances in science. Considering that Paley was politically well connected and not a man anyone would want as an enemy, one can imagine Erasmus Darwin's reluctance to promote evolutionary theory. In the final chapter I will consider in detail how such teleological philosophies can be harmful to religion in an age where science is all-embracing. Briefly, scientific studies *do* begin with a blank slate free of opinion or belief, and so their results are reliable. This makes the teleological argument the irrational antagonist in the debate with the rational (science). That's not my opinion—that's simply laying the facts on the table. Even Charles Darwin felt compelled to point this out in his books, and *he* began with a belief in God. No, it is not good for religion to defend God with logic so obviously amiss. And, since in doing so one is exploiting people's genuine beliefs while knowing that the argument doesn't hold up, it's also immoral.

One of Erasmus Darwin's failings over evolutionary theory was his belief that animals acquired characteristics during their lifetimes. This idea was taken up more spectacularly by Jean-Baptiste Lamarck, botanist to the French king before the Revolution, and professor of zoology at the National Museum of Natural History in Paris in the aftermath. Alas, the false philosophy was even named after him—it became known as Lamarckism. Lamarck's most famous example related to the giraffe, which he claimed had strained its neck to reach the prime leaves at the tops of trees, so that over successive generations the neck had become long. A charming version of events reminiscent of Kipling's "Just So" stories, but no more true.

Erasmus Darwin and Lamarck, although heroes within scientific and philosophical circles in Britain and France, never had much of an impact on society in general. And then, although recognizing the idea of evolution, in accepting that animals and plants change over time, neither came up with the right idea of how this might happen. Evolution was something of a mystery, and as such remained a subject for science. Now for the small red book.

Henry Gosse (full name Philip Henry Gosse) was born in Worcester, England, in 1810, and led a career as a naturalist and popularizer of natural science; he also virtually invented the seawater aquarium and contributed important work to marine biology. At the age of seventeen Gosse sailed to Newfoundland to serve as a clerk, but quickly became a self-taught entomologist. After five years in Canada he experienced a religious conversion. From that moment, Gosse would see all things natural as God's creations. He became blinded by faith, although this did not affect much of his biological work, such as his detailed descriptions of animal anatomy and behavior. It did, nonetheless, affect his opinion of Charles Lyell's work on the vastness of geological time.

After a brief stay in Alabama, Gosse returned to England in 1839. In 1844 he set sail on a two-year trip to Jamaica, then returned to London, and finally settled on the Devon coast to become the archetypal gentleman scientist. He maintained close links with the church, married Emily Bowes, a devout Christian, and lived very happily, having a single son (Edmund, who would eclipse Henry as an author).

Gosse published many books over the course of his career. The library of the Natural History Museum in London still has shelves filled with multiple copies of his first and subsequent editions, indicative of Gosse's standing and influence in Victorian natural history circles. But he is perhaps best known for his small red book—*Omphalos*, subtitled *An Attempt to Untie the Geological Knot*. Omphalos is Greek for navel: The reference is to Adam's navel and the book addresses the question whether, as the first human (according to the Bible), he had one at all.

Omphalos was the first book written by Gosse after his wife's death. In it Gosse set forth his theory that Adam did not require a navel because he was never born, yet would have had one all the same. By the same logic, the first trees created would have contained growth rings even though they had not grown. Applied to the fossil record, which had by now grown to such a size that it could not be denied, the theory stated that fossils were forms instantly fashioned by God at the moment of the creation, and had never been living things. Gosse thought that he had cracked it—he had reconciled the geological evidence, which had seemed to be chipping away at the Bible's authority, with God's activity. But, unlike his other books, *Omphalos* was a failure, both intellectually and financially.

A more cynical interpretation of Gosse's theory was that God had created an enormous and unnecessary lie for all humankind. This became the accepted view, and Gosse was perceived guilty of crimes against science. In this view, God had apparently hidden fossils in the rocks to tempt geologists to faithlessness, a ridiculous argument sadly still wheeled out by some creationists today. Even Gosse's friends were taken by surprise. Gosse tried repackaging his book with a different title—*Creation*—in case it was that which had held it back. It was not, and sales did not improve.

Gosse was distraught. According to his son, he "closed the doors on himself forever." To rub salt into his wounds, in 1859, before the dust never actually kicked up by *Omphalos* would have theoretically settled, the first of the green books I mentioned (at the beginning of this chapter) landed on Gosse's desk. It landed with a thud. It contained the right answer—evolution by natural selection. The fossils were actually preserved ancestors of modern species. Let's move on to that bombshell.

Darwin and the Tree of Life

Charles Darwin was a grandson of Erasmus Darwin. Born in 1809 in Shrewsbury, England, near the Welsh border, he did not seem destined for great things. His academic career did not start well: His father, a well-respected Shropshire doctor, assessed Charles as "ordinary," and considered the best he could aim for was a career as a clergyman.

A life in the church actually appealed to Charles. He had become fascinated by beetles, and entomology was a common hobby among clergymen. But it became more than a hobby for Darwin, who would allow it to distract him from his university studies at Cambridge. He made the grades, but failed to shine.

While at Cambridge Darwin had the good fortune to work under the respected geologist Adam Sedgwick. Sedgwick said something to Darwin that would change the way he thought. During a field trip to a gravel pit in the Midlands, Sedgwick found a large fossil shell, but claimed that it must have been thrown there, because it did not belong to rocks of such age. He continued, "It would be the greatest misfortune to geology as it would overthrow all we know about the superficial deposits of the Midland counties." Darwin could not understand why a man of science would be anything but delighted by a new piece of evidence just because it did not fit with an accepted group of facts. He primed himself for radical thought.

Post-Cambridge, Darwin's imagination set sail after absorbing the travel tales of the German explorer Alexander von Humboldt in South America. He commented, "It strikes me, that all our knowledge about the structure of our Earth is very much like what an old hen would know of the hundred-acre field in a corner of which she is scratching." Darwin was in no mood to stay in the corner of his field. In 1831 he took ship himself on the *Beagle*.

The *Beagle*'s voyage around the world was planned to take two years, but lasted five. Darwin was lucky to be on board at all, since his father had refused him permission; the impasse had only been broken by the intervention of Charles' uncle, the famous potter and innovator Josiah Wedgwood, who intervened in his favor. The captain of the *Beagle* was Robert Fitzroy, a deeply religious young man of twenty-six, who had invited a scientist along to find evidence to support the literal truth of Genesis, an activity proving popular at the time. Darwin himself questioned neither the fact nor the relevance of biblical teachings. But in hindsight it became only too clear that Fitzroy had chosen the wrong man for the job; later the guilt would become unbearable and he took his own life.

The *Beagle* was a small ship, and for the first couple of weeks of the journey Darwin was hopelessly seasick. Then, when the expedition docked at Rio de Janeiro, Darwin encountered his first tropical rain forest . . . and at once he felt overwhelmed. He wrote to his scientific mentor that he feared he was not the man to bring home the right specimens. But maybe it was his limited scientific training that allowed his mind to roam free—free enough to observe what was actually out there, rather than what fitted with accepted theories.

In South America, Darwin found the skull of an extinct animal related to a rhinoceros, and a jawbone the size of an elephant's that actually came from an extinct rodent related to the sloth. Darwin thought it curious that God should place such huge ancient creatures in the home of their tinier kinds today. Fitzroy explained that the fossils were of two creatures that hadn't made it to the ark in time. Darwin began to wonder whether the animals he saw alive in the rain forests might be descendants of the ancient beasts.

Maybe the rain forests held not just too many trees but too much flora and fauna for Darwin to see the proverbial woods. Maybe it took the more sparsely populated Galapagos Islands to compose Darwin's thoughts. Darwin was a meticulous note-taker, and if he wanted to make a comprehensive study of any environment the Galapagos would give him a better chance.

Darwin listened to local accounts of the wildlife. One fateful day in 1835, Vice-Governor Larson happened to mention that there were subtle differences between the giant tortoises of the different islands, adding that he could quickly tell which island any had come from. Darwin took a ride on one of them for a while, before heading to the coast. There he took an iguana by the tail and threw it into the sea. It immediately returned to the land, as if it were not made to survive the predatory sea lions. Finally, he observed for himself the differences in beak shape between the local mockingbirds—the three species of three different islands—and the penny dropped.

The Galapagos Islands are the peaks of submarine volcanoes, which have never had a land connection to a continent. All of their fauna and flora got there originally by colonization over the sea from South America. But Darwin knew of only one species of mockingbird on that continent, while he observed three in the Galapagos. Where, he asked, did the other two suddenly appear from?

In one of Darwin's earlier notebooks exists the first-ever figure of an evolutionary tree, sketched rather primitively yet bearing a message as plain as day. Charles had converted Erasmus Darwin's artistic musings into a stark, brutal diagram that was well worth a thousand words. In fact there were just two words above it—"I think."

In this diagram one gets a glimpse of the myriad questions massaging Darwin's mind as to why different animals and plants share certain characteristics—the feathers of all birds at the higher level, the talons of all birds of prey at a lower level. The answer he arrived at is that each of these groups shares a common ancestor. All birds evolved from a single ancestor that lived before birds existed, and all birds of prey evolved from a single bird ancestor that lived early in the history of birds. The further back in time the common ancestor existed, the larger the group it leaves behind today. *This* is the tree of life—and Darwin had been led to it by his questioning of geological isolation and shared characteristics. A single colonization by a South American mockingbird had given rise, by branching descent, to three different species on Galapagos. But *why* did the characteristics change? Why did the Galapagos mockingbirds have different beak shapes?

The key to Darwin's solution to this new problem was his recognition of *function*: that each characteristic was suited or adapted to a certain task. Each feature—the feather of a bird, or the talons of a bird of prey, to which Richard Owen had so painstakingly drawn attention—served a purpose that helped the animal to survive in its specific environment or lifestyle. The Galapagos mockingbirds' food was different on each island, and so a different-shaped beak would be favored in each local environment. Darwin put this idea together with two others—that the individuals making up a single *species* also show *slight* variations in their characteristics (just think of humans); and that more individuals are born than survive.

During his voyage on the *Beagle*, Darwin had witnessed nature at its most cruel. Just as in the apparently more tranquil English countryside, but more shockingly, encounters between predator and prey in South America or Africa can be vivid, violent, and cruel; and in both environments they are natural. A lion locks its jaws on a zebra's throat just as swallows pick flies from the summer skies: Both are acts of predation, but in the former the blood and guts are visible. This attention-grabbing spectacle of predation offers a more emotionally charged view of the process, raising one's awareness of its sheer scale. Predation, in such a scene, becomes an unmistakable reason why more individuals of a species are born than survive. Now, Darwin thought, what if one variation on a characteristic improved an individual's chances of survival? In the case of the zebra, a color pattern that is confusing to the eye would be useful. The lion, on the other hand, could conceivably benefit from stronger jaws to hold on to its prey better. Maybe the characteristics that aid survival are those that are passed on most readily to the next generation, thought Darwin. But how?

This is where Darwin departed insightfully from Lamarckism. The individuals that possess the characteristics that best aid survival—the "positive" characteristics—are, by definition, more likely to live the longest within a species. If they live longer, they will reproduce more often and produce more offspring. Now, if the characteristics are passed on to the next generation during the process of *reproduction*, then it stands to reason that, with the passage of enough generations, the positive characteristics will win out and flood the population.

The characteristics that show variation within a population today may be rather secondary ones. But, Darwin postulated, what if the same mechanism applied to primary characteristics, too? Consider if feathers helped a lizard to catch flying insects. Then a population of a lizard species that included feathers in some individuals would, over generations, become a population of exclusively feathered individuals. The feathered lizards would no longer mate with nonfeathered varieties in other areas, since they appear so different, and a new species would be born. This mechanism Darwin called "natural selection." It was a mechanism by which evolution—the branching of the tree of life—worked. Why do animals share characteristics? Because they have a common ancestor, asserted Darwin. Natural selection involved common ancestry and descent with modification, and in the struggle for life favored races have been preserved.

That last phrase I have adapted from the subtitle of the book that landed on Gosse's desk in 1859—*On the Origin of Species*. This was the first book to explain how evolution happened, finally making the theory ripe for the public domain. Its pièce de résistance was the lucid and graspable evolutionary tree. *On the Origin of Species* was about the same general size as Gosse's *Omphalos*, although about half as thick again. It brought Gosse's world tumbling down.

A couple of years ago I found, tucked inside one of Gosse's many books in that antiquarian bookshop back in London, a letter from Gosse to his insect-cabinet maker, dated July 30, 1878, from his home near Torquay in Devon. Clearly, Gosse spoke and thought as he wrote in his books, for the prose in the letter was charming and elegant, with an affectionate account of a precisely finished insect-cabinet to match his famed lyrical descriptions of the natural world. Yet something struck me in the closing paragraphs. The cabinet maker had enclosed some specimens of moths in the drawers of the cabinet as a gift, and Gosse replied, succinctly, that he had not seen them before and that "they are very closely alike."

Gosse had used a fair amount of ink to describe his joy over a wooden and glass cabinet, yet was extremely short on a subject that should have interested him most: natural history specimens. It is apparent that the insect-cabinet maker, someone surely not averse to science, had made a reference to Darwinian natural selection, and how it might explain the similarity between his moths. The moths, he doubtless considered, were closely related. Gosse likely spoke with emphasis—"they *are* very closely alike"—reluctantly agreeing with the cabinetmaker's implications. The likeness supported the theory of natural selection—the moths were probably closely related. This is perhaps indicative that, some thirty years after *Omphalos* and *On the Origin of Species*, Gosse had lost his personal battle with evolution.

Before Charles Darwin, most people in England lived in farming communities and did not venture far from where they were born. Change was not a concept they associated with. People occupied the family home, even the family trade, for generations. They never doubted the family Bible. All was extremely stable.

Gosse came along and provided books that drew out the beauty in the animals of the countryside they were accustomed to. *God's* beautiful creations. The books were most welcome, for they improved the quality of people's lives. Stephen Jay Gould described Gosse as the (famed television naturalist) David Attenborough of his day.

The Industrial Revolution and democratic reform had made their appearance, and perhaps primed people for change. But these events filtered into society gradually. Darwin's theory, like his book on Gosse's desk, would make its entry with a thud. It is well documented that the *Origin* came to Gosse as a stake through his heart. He would never recover, as the case of the similar-looking moths attests. He prayed that he might not taste death but meet Christ in the air at his Second Coming, and he was bitterly disappointed when he realized, in his final days, that he would die like everyone else.

Gosse had corresponded amiably with Darwin on the subject of orchids. Maybe his conscience cried that he had been wrong over God's fossils. Maybe he too experienced the intensity of intellectual feeling that accompanied the recognition of natural selection.

Gosse, nevertheless, found the energy to argue that the mechanisms of evolution are insufficient to explain the living world in total, and concluded that, because of all the insufficiencies in our understanding of the mechanism, evolution did not happen at all—

just as creationists do today. Each time someone proposes a different view on the accepted mechanisms (most of which are explanations for new observations), creationists pop up to say that this refutes evidence for evolution altogether. This is not the way to make intellectual progress. Fortunately, Gosse continued to find time to reveal the romance in natural history, and to publish it for the Victorian fireside.

Before Darwin, the general population had religious ideas ingrained in their minds. Species were not linked within a family tree—they had existed unchanged since the moment of their creation. There was no doubt (except within certain philosophical circles) that God had created every plant and animal species, including humankind. Imagine how, in the 1860s, people were suddenly asked to change their outlook from this to one in which animals and plants evolved from a single-celled common ancestor, in a system that seemed not to require God. Worse than that, how would they now find solace in the world? With no substantial answer to this question, religion survived.

Having now established the principle that life evolved through natural selection, we are confronted with what may at first seem an obvious question—if life evolved, what did it evolve from? Or, to put it another way, what was the first life form on earth? We must return to the starting point of this chapter. Is it correct to say, as the account in Genesis does, that the first forms of life to appear after the creation of the seas were plants? To answer that question we must look at the mechanism by which evolution happens.

But first, an important point of business: Although religion did survive the early Victorian era, it continued in a more defensive and antagonistic manner—it met evolution head-on, rather than trying to work with it. We will consider in the final chapter whether perhaps religion did not need to take up such a confrontational stance.

Maybe it could have worked *with* evolution, and maybe if it had done so it might have been strengthened—strengthened by removing some of the contradictions or inconsistencies in scripture that had always, secretly, provided the individual with cause to doubt.

At this point I'm feeling terribly guilty—guilty of something I complain that others (including other biologists) so often do. At the beginning of this chapter I did say that there were *two* green books. I have neglected Alfred Russel Wallace. In the 1850s, as is evident from Darwin's letters to Wallace (Wallace's replies have been lost), both Darwin *and Wallace* were independently thinking along the same lines. Although natural selection has come to be known in shorthand as "Darwin's theory," the theory was born separately in *both* their minds—they were *cofounders* of the idea.

Alfred Russel Wallace was born in Monmouthshire, Wales, in 1823. *His* green book—*The Malay Archipelago*—did not appear until ten years after Darwin's, in 1869. This was his first book to discuss his ideas of natural selection, although it was followed by many others and had been preceded by many contributions to scientific journals. The most significant of these was a paper of 1858 which was read, together with one by Darwin, to a meeting of the Linnaean Society in London—marking, in effect, the copublication by the two men of the theory of evolution by natural selection, albeit on this occasion addressed strictly to the scientific community.

When Wallace was five, his family moved to Hertford, north of London. After leaving school early, Alfred worked as an apprentice builder, then as a surveyor in London. After a brief period of unemployment he was hired as a master at the Collegiate School in Leicester. Here he met Henry Bates, whose fascination in beetles rubbed off on him. The two men went on to collect insects together, and became interested—more so than Charles Darwin—in the ideas of evolution as expressed in Erasmus Darwin's *Temple of Nature*. Cutting to the chase, in 1848 they set sail together, on the *Mischief*, for the Amazon—the mecca for insects.

After about a year in entomological heaven, the two naturalists went their own ways—Wallace explored for three further years before returning home, while Bates ventured even deeper into the rain forest, where he would stay for another nine years. During this spell Bates discovered the concept of mimicry, by which harmless species would *evolve* the color patterns of unpalatable species to avoid predation. Mimicry would later provide supreme independent support for the theory of natural selection. Wallace, meanwhile, would plainly and simply form the pieces of the natural selection puzzle in his mind—as inaudibly and unassumingly as that.

Returning to England, Wallace found his sea legs restless, and after a year and a half he went away again, this time to Southeast Asia. From studying the local wildlife he realized that a line could be drawn through the area separating very different faunas, as widely divergent as in any two parts of the world. This became known as "Wallace's Line." Then, in February 1858, in Sarawak (Borneo, Indonesia), in the midst of a severe attack of fever, he put the pieces of the evolutionary puzzle together. "There suddenly flashed upon me the *idea* of the survival of the fittest," he wrote. Over the following three days he developed his idea into a letter and sent it to an acquaintance he knew also to have an interest in evolution: Charles Darwin. This prompted Darwin—with a sharper sense of the stakes—to publish the defining book; and the rest is history.

Most of the personalities involved in the history of science in the story so far were people who did not need to think about a salary. They were generally well off. Wallace, on the other hand, was not. Darwin was afforded the friendship and respect of the middle classes, but Wallace's family, although originally of middle-class origin, fell upon hard times and all were condemned to the insufferable (in the eyes of the scientific establishment) fate of having to work for a living. As a result, Wallace had few friends or champions when—and where—it mattered. Regrettably, Darwin and Wallace's theory of natural selection, as indeed it was known in its infancy, became shortened to "Darwin's theory." Wherever I repeat the sin in this book, please remember what I really mean.

Darwin's religious stance varied over his lifetime. Over time he grew to doubt the Old Testament, although this trend is somewhat exaggerated in many modern accounts, which can be guilty of putting words into his mouth. It seems that Darwin was rather more agnostic than atheist, although he did have spells where he became increasingly doubtful of God's existence. Either way, he did rather shy away from the subject in public. Wallace, on the other hand, remained religious, and was not afraid to let people know it.

Up until the nineteenth century, science had always been mindful of the dangers of contradicting the church. Suddenly the balance shifted, and Wallace experienced the opposite danger: That his scientific legacy might be eclipsed because of his adherence to the Bible. Wallace felt able to reconcile his science and religion. While critical of those resting on blind faith—those who sought to destroy evolutionary theory—Wallace's faith was strengthened by intelligent conviction. He would turn to writing on this subject, but, as the curtain fell on his illustrious career, the scientific establishment would unfairly mute his words.

So, we have finally covered the theory of evolution by natural selection. We now know that any random change to an individual that helps it to survive better in its precise surroundings will increase its reproductive potential. Accordingly, the genes that code for this change will be passed on, and all progenies will carry this new adaptation. Ah yes, the gene: one of those things that Darwin and Wallace didn't know about, and were forced to work around.

They adjusted rather well, though, and all but predicted what was to come.

Now the final part of the jigsaw puzzle: In 1865 an Austrian monk named Gregor Mendel published a paper in an obscure journal describing in detail how characteristics are inherited through generations. Mendel crossed different breeds of plants in an attempt to work out the nature of heredity. He realized that many characteristics, such as the color of pea seeds, are passed down in pairs of discrete units. In 1909, when this work was rediscovered, these units would be termed *genes*. It is possible that Mendel's paper even languished, unread, in Darwin's study. Darwin and Wallace remained blinded by the alternative, incorrect explanation that the characteristics of parents are blended in their offspring—in their view, a black mouse mated with a white mouse should give birth to gray mice. Mendel predicted they would produce some black mice and some white ones. Unfortunately Mendel did not live to see the worldwide recognition his work would eventually receive.

When the British researcher Francis Crick and his American colleague James Watson unravelled the structure of DNA (*D*eoxyribo*N*ucleic *A*cid) in 1953, the way information flowed between generations became apparent. Their double helix ("twisted ladder") model explained how the genetic information for making a body could be faithfully copied when cells divided: through dividing itself and ensuring replica copies of each half. It also explained how cells could carry a code to build the machinery and structure of a cell (the enzymes and other molecules): The rungs of the DNA ladder were made of pairs of chemical bases that carried messages in linear code.

If the copy of the new DNA in an offspring was not exact—if errors occurred during the replication process—then the change in this instruction manual might lead to a different structure on the animal's body. We see only those changes that are advantageous to the animal—individuals with other, negative changes cannot compete and their "mutations" become extinct. This is how animals on the earth today came to be so well adapted to their environment. This all follows Darwin and Wallace's logic, and as a result provides staunch support for their theory.

Even more recently, we have moved from simply knowing the genetic blueprints to understanding how they are translated into a living body—a modern subject known as evolutionary developmental biology, or "EvoDevo." Here, a developing embryo is considered in detail as a three-dimensional ball of cells, where each cell has a coordinate in space. Although each cell carries the same genes as all the others, a change in one cell triggers a cascade of changes through neighboring cells, and throughout the development process, genes for different characteristics become activated at different times. Genes for the features of the head are activated in a frontal region, and those for a back in the top region, for instance. Just to confirm these findings, researchers can perform clever tricks, such as activating the genes for an eye in the part of a fly embryo that will become a leg. The adult fly appears with, right on cue, an eye on its leg.

EvoDevo can explain that sometimes gradual links between related species—those "missing links" to which attention is drawn by creationists—are not necessary. Consider a zebra embryo, and the region of cells that will develop into the skin of its back. Alternate cells will produce a black pigment, while those in between will remain white. If the genes for black pigment are activated while there are 64 cells in this region, then the adult will have 32 black and 32 white stripes. If cell division is allowed to continue for one more step, so that there are 128 cells in this region when the pigment genes are activated, then the adult will have 64 black and 64 white stripes. As a result of this process, the most closely related species of zebra will have 64 or 128 black and white stripes, but not values in between. A zebra with 76 or 92 black and white stripes—the so-called missing links—is not a possibility.

From Fireworks to Photosynthesis

Up to this point we have covered, through the vehicle of the history of science, the basics of evolution, including one of the central mechanisms by which it works, natural selection. There are other mechanisms, including sexual selection—which centers on improving attractiveness to a potential mate—but by now we have a feel for the robustness of the science. We know that life diversified through the various mechanisms of evolution: Life evolved. All life can be traced back to a single, common ancestor—a single cell. But what about the next step back? How did that *first* cell originate?

Remember that, in the earth's early, lifeless history, the rocks and the seas contained atoms joined together in the form of molecules—molecules of all different kinds. The atomic building blocks were themselves a product of the Big Bang—the origin of the universe—when all the energy in the universe came to be. There was finite energy on the earth, and some of that now existed in the form of matter, made of atoms and simple molecules. The next stage in life's prehistory involved simple molecules joining together to form complex ones.

Remember that at this point in the earth's history the tree of life was a thing of the future. The simple molecules would join together, forming complex molecules that would in turn form the seed from which the tree of life would grow, via the processes of evolution. The animals and plants of today can be traced back to their ancestors, say in the ice age. The sloth can be traced back to the giant ground sloth, for instance. But if we continue to project back in this manner, sloths will eventually be traced back to fishes, then to sponges (before the animals with backbones had evolved, or branched from the tree), and finally to a single-celled form. It is the formation of this single cell—the first living entity—that we will explore now.

In 2008 Professor Jeffrey Bada of the University of California, San Diego, decided to open several small cardboard boxes he found among the usual dusty junk piled up in the corner of his lab. The boxes, with some accompanying lab books, belonged to his former mentor Stanley Miller. The Jewish American chemist and biologist Miller was just twenty-two years old when, in the 1950s, as a PhD student at the University of Chicago, he conducted a famous experiment.

Bada broke open the tape and found inside the boxes some two hundred small glass vials containing dried samples of . . . something. To discover what these samples really were, Bada and his team subjected them to chemical analyses to reveal their spectrum of elements. The results showed that the vials contained twenty-two amino acids. That was important to know.

Amino acids are complex molecules that themselves form the building blocks of proteins: Proteins are chains of amino acids. They are critical to life. Twenty different amino acids are found within proteins, and these catalyze at least most reactions within living cells, as well as controlling virtually all cellular processes. The precise amino acid content of a protein molecule, and the sequence of those amino acids within it, are determined by the sequence of "base" molecules in the gene that encodes it. DNA contains a code to make amino acids and, ultimately, proteins. But amino acids don't simply assemble by themselves, even if all the constituent atoms are present. If we can work out how the first amino acid molecules formed on the earth, we can move on to the less challenging question of how the amino acids came together to form proteins, and then to ask how the proteins formed cells. It is in answering that first conundrum that Stanley Miller's glass vials are important.

Miller's famous experiment in the 1950s tested the idea—expounded by, among others, the Russian biochemist Aleksandr Oparin—that life evolved in primeval seas through the self-assembly of the simple molecules that existed then. In a glass laboratory flask, Miller mixed water with methane, ammonia, and hydrogen—the kinds of gases thought to have dominated the earth's oxygen-free atmosphere some 4,000 million years ago. As expected, nothing happened . . . until he introduced electric sparks to simulate the profuse lightning strikes of that primal period. The sparks turned the mixture in the flask red, then yellow-brown. Five amino acids had been created, including glycine and alanine—common constituents of proteins.

Miller became an instant celebrity, although the newspapers, true to form, overstated the case. They misunderstood what he had done, asserting that he had recreated life in his lab. He had not. But he *had* formed amino acids from the conditions found on the early earth, albeit only a quarter of those that life possessed. This experiment acquired iconic status for simulating pre-biotic chemistry.

Soon after, Miller revised his experiment by injecting hot steam into the glass flask so that conditions resembled those found in an erupting volcano. But Miller published only the sketchiest of details of this work, and the experiment could not be repeated by others later because the unknown apparatus Miller had used had been lost. Lost, that is, until Jeffrey Bada opened those cardboard boxes. For what they contained were the glass vials from Miller's second experiments, and what *they* contained was the full complement of twenty-two amino acids. He had done it! Miller had created the entire range of amino acids found in life, and in much higher yields than in his first experiments. Unbeknown to himself, and to the world until recently, his second mixture appeared to be much more fertile.

Today, most volcanoes belch out gases such as those used in Miller's second experiments, and eruptions are almost always accompanied by electrical storms. We believe that the early earth was highly volcanic, and so it appears likely that those early volcanoes were pre-biotic factories, generating amino acids that were washed into pools or coastal bays, where they amassed. A pre-biotic soup—a cradle for life—had been prepared. Miller had demonstrated that God was probably not necessary to make life on earth.

Of course, another step is needed for life to emerge: Those amino acids must come together to form proteins and then cells. The formation of proteins is the easy bit. Amino acids need no encouragement to simply join together, at the expense of losing hydrogen and oxygen atoms in the form of water (a reaction known as dehydration synthesis). A cell, on the other hand, is no mean structure, even though the first cells on earth were simple in comparison with many today. Each animal cell contains power plants called mitochondria, an organizing center called a nucleus (containing the DNA), a border patrol in the form of a double-layered membrane that is choosy about what it lets in and out, and a whole bunch of activities taking place within the outer membrane's boundary. The first cells, in contrast, were just spherical membranes, each with a strip of genetic instructions floating on the inside. Still, they needed the component parts to come together at precisely the same place, although it is believed that this happened in stages and not all at the same time.

The outer membrane is made up of fatty acids, which can be made from the breakdown of amino acids into smaller units. Fatty acids in the membrane have "hydrocarbon tails," which can form from the hydrogen and carbon monoxide gases also found at volcanic sites. Maybe the primeval pools and coastal bays also became filled with fatty acids. It is well known that, in high concentrations, fatty-acid molecules spontaneously arrange themselves to form the membranes of empty vesicles. These membranes are extremely stable but dynamic, in that the molecules flip over to invert, and then right, themselves. In this way, they can bring other molecules in the water into the vesicles, so that they are no longer empty.

Some of those other molecules present in the water on the early earth were RNA (ribonucleic acid) units or "nucleotides"—a simpler version of DNA nucleotides, since they form a single strand rather than a double one. Some sequences of RNA nucleotides are more stable than others, and these, once carried inside the fatty acid vesicles, could begin to activate their code for arranging amino acids into specific sequences to produce certain proteins.

I could continue, but I have said enough to make clear the basic premise here: Namely, that it is theoretically possible, and now all but demonstrable, that a proto-cell could spontaneously generate under the right conditions and with the appropriate molecular building blocks. Such conditions and chemicals are those believed to have existed on the early earth, in volcanic regions that included hydrothermal vents, thousands of meters under the sea. We also know that RNA is capable of replicating itself, and so that very first cell held the ability to divide and duplicate, leading to a population of a species. This is how life began, around 3,900 million years ago.

This brings us back to Erasmus Darwin, who predicted how life might have formed within a primeval ocean. I mentioned that, as a poet, Erasmus Darwin influenced Percy Bysshe Shelley, who of course became a far more famous poet. In company with Erasmus Darwin, Percy's wife, Mary, would listen intently to Erasmus Darwin's good logic. She applied this, in her imagination, to humans, and produced a piece of fiction that would employ precisely the concept of the spontaneous generation of life in a dead human body, electricity and all. The book was *Frankenstein*. In her story, just as in the original emergence of life on earth, given energy, the right molecules can be induced to aggregate into something that is self-perpetuating. From here, evolution by natural selection can take over.

Through errors in the process of replication of its RNA, that first species evolved into new species. Eventually, three branches had formed on the primitive tree of life—bacteria, archaebacteria, and eukaryotes. All three are single-celled forms, distinguished by different internal chemistries and by components to their cells that are made differently. The eukaryotes involve more complex cells, and will be the focus of the following chapters. But before the eukaryotes, the simple-celled bacteria and archaebacteria had evolved. The first species of these obtained the energy they needed to undergo their cellular functions by breaking down molecules that existed in the water. Again, it all comes down to energy. Clearly this system is not self-sustaining. As I said, there is finite energy on the earth. To avoid rapid extinction, life needed extraterrestrial energy. It got it from the sun.

An early bacterium evolved a molecule that could use the energy in the sun's rays to gain electrons from water—it could *photosynthesize*. The "cyanobacteria" were born. Today, we find fossils of cyanobacteria as much as 3,600 million years old, so they must have evolved very near to the beginning of life on earth.

Finally we have found out how life was possible on earth, and how the first cell could evolve into species capable of extracting extraterrestrial energy. But what has this got to do with plants—"grass," "herb," and "fruit tree"? Why did the author of Genesis use these very words?

Looking at it from another angle, after exposing the scientific story, what's so special about the evolution of the first photosynthetic cells, the cyanobacteria? Should *they* appear on the first page of Genesis, as the first representatives of life of any kind?

Well, the evolution of the cyanobacteria was one of *the* monumental stages in the history of life on earth. First, the initial life forms were on a collision course with extinction, since they were gradually using up the earth's finite energy supplies. But the cyanobacteria evolved and brought new energy to the earth—energy from the sun. Then, in order to make their food, the cyanobacteria extracted not only water from their environment, but also carbon dioxide. In return they gave the atmosphere oxygen. An oxygenated atmosphere is far more conducive to life, and the protective ozone layer that formed guarded the earth's surface from the sun's high-energy ultraviolet rays, which destroy biological material. The cyanobacteria primed planet earth for the next stage in life's history. They were absolutely necessary.

Having said that, the answer to the second question, about the appropriateness of cyanobacteria to the creation account of Genesis 1, is that they could not be more fitting to represent the first stages of life. *But* cyanobacteria are nowhere to be seen on the Bible's first page. Instead, as the first forms of life, we have "grass," "herb," and "fruit tree." So why did the author pen these particular words?

The ancient Israelites would have been oblivious to any single-celled life form, let alone cyanobacteria specifically. The writer of Genesis could not have had a word for cyanobacteria, or anything remotely similar, in his vocabulary. Single-celled photosynthetic forms—some of *the* earliest life forms on earth—would have lain well beyond his grasp. But could he have been translating onto paper a message that somehow inferred a primitive form of life—a form that photosynthesizes?

Photosynthesis, of course, is synonymous with plant life—plants, as we all know, are the life forms that make their food from water, carbon dioxide, and sunlight. Plants are members of the eukaryote group that I mentioned would be covered in the following chapter. They are higher life forms than bacteria. Yet the fact that those early cyanobacteria could photosynthesize like plants today is no accident. Cyanobacteria became incorporated into eukaryote cells to form organelles called chloroplasts—the photosynthetic factories of plants. So the plants mentioned in the third stage of the Genesis creation account actually contain modified "cyanobacteria."

It is extremely appropriate that the author of Genesis chose plant life for this third stage in his creation account—it accords particularly well with the science. In fact he couldn't have chosen better. But he was not to know this. Why he so appropriately listed grass, herb, and fruit tree is a question we will leave until later. It is also interesting that water was absolutely necessary for life to have appeared at all, and water was singled out in the Genesis account's previous stage. Indeed, today, when "astrobiologists" search for life on other planets, they first search for signs of water.

Whatever the case, these primitive, photosynthetic forms—the cyanobacteria—represent a monumentally significant stage in the history of life because they oxygenated the atmosphere. Were it not for them, we would not be here today, although I could possibly make a similar statement in each chapter from this point on. So we have now seen science and Genesis agree once more—from light, to the creation of seas, to the first forms of life, plantlike organisms; but we still have a way to go before the end of our journey. Primed with an understanding of evolution, it is time for the next stage in the creation account of Genesis 1, and in our scientific understanding of the history of life on earth. Time for the multicelled animals.

Sight

"Let there be lights in the firmament of the heaven to divide the day from the night, and let them be for signs and for seasons, and for days and years."

The author of Genesis recorded that "God made two great lights: the greater light to rule the day, and the lesser light to rule the night." What was he trying to say here? He had already covered the appearance of the sun, which illuminates the earth's surface for part of the day, and effectively disappears for the rest. This fourth phrase in the creation account cannot be referring to that, although it appears to invoke the sun again, somehow. In fact this is the phrase in Genesis that has most frustrated biblical scholars.

If the scripture is not immediately obvious, we should turn our attention to the science, and consider what happened next in the history of life on earth. Then we might look for any overlaps with this phrase from the first page of Genesis. As a sign of things to come, there was indeed an introduction of light somewhere in the evolutionary history of animals, and this was a monumental step that changed life forever—worthy of a few words on the first page of the Bible, perhaps?

Let's turn back again, and not for the last time, to T. E. Lawrence. I came across another volume in that antiquarian bookshop in London that grabbed my attention, a copy of Anna Sewell's *Black Beauty* that contained an inscription by Lawrence dated 1934. The message, to a member of the Yarwood family of Northwich, discussed Lawrence's attitude to animals and to motorbikes. "Unfortunately I went off on machines," he wrote. "Machines are extensions of our faculties." In other words, add an engine and wheels and we can move faster. But what if we consider a new body part—a natural extension of our faculties—rather than a machine? What if we evolved a new body part that conferred some advantage on us, in the manner of a motorbike?

First, let's remember the energy lesson: The improvement afforded by the new body part must outweigh what it costs us in energy to make the part as our body develops, and then to maintain it. So that part must be *really* useful to us, particularly if it is sizable and carries a high cost—a new, faster engine in a lion, for instance. And if it does pass the test of natural selection, then our neighbors had better look out: In the case of the lion, the zebra and wildebeest on the Serengeti would become increasingly nervous. But immediately natural selection would begin to act on them, too. Such a line of reasoning will emerge as a theme of this chapter. What would happen to those zebra and wildebeest will be a theme of the next.

"My bike does 96 and goes on for as long as my strength lasts," Lawrence continued in his inscription. That was poignant, since not much more than a year later he would have a fatal crash. At the end of this chapter I will consider the extra power afforded to living creatures by the biological equivalent of Lawrence's motorbike in life's history—that is, the eye.

Now for an apparently surreal statement that will, later, link up with Lawrence's phrase about the "extension of our faculties"—that *the world we experience and know does not really exist*. This might appear just as confusing as the words of the author of Genesis above, but it's a scientific statement. By the end of this chapter, this statement will appear not as surreal but as very real; and the reason why can help to make sense of the second reference to the sun in Genesis.

Back to the history of life.

So far, single-celled forms had evolved that oxygenated our atmosphere and provided a protective ozone layer. And the earth and its climate became even more suitable for life—it stabilized. Mars and Venus also appear to have started out wet, but they lost their oceans. The earth, on the other hand, has a natural thermostat to regulate its temperature. It's all to do with cycles.

Volcanoes contribute carbon dioxide to the atmosphere, which keeps the planet warm through the greenhouse effect. The warmth causes seawater to evaporate and form clouds, and then it rains. The rain contains dissolved carbon dioxide that makes it slightly acidic, so that when it interacts with the surface rocks, carbon-containing minerals are dissolved into rivers and wash out to sea. The minerals precipitate out and form new carbon-containing rocks. Plate tectonics move these rocks into the molten mantle and further volcanic activity brings the carbon, as carbon dioxide, back into the atmosphere—and the cycle starts again. The rainfall can vary, and acts as a thermostat by controlling the proportion of greenhouse gases in the atmosphere. When the planet is warm, rainfall increases, speeding the rate of atmospheric carbon dioxide removal and so cooling the planet. When it is cold, rainfall decreases, allowing volcanic gases to build up in the atmosphere, thereby warming the planet.

Now what were those single-celled life forms that exploited this perfect planet?

MICROSCOPES AND THE MOLECULAR CLOCK

An assertion commonly found in books and web sites is: *"Of the five senses, the most important is sight."* Certainly, vision is the sense we employ when we classify life in biology. But our eyes are adequate only up to a certain point, beyond which the unaided eye fails to offer us any help. When life gets very small, we become unaware of it. How we came to know about the single-celled bacteria, archaebacteria, and eukaryotes is fascinating, but I am also making a point about the importance of sight. The reasons are manifold and they show up in the single-celled bacteria story and many others throughout this book. As we shall see.

Although the glass lens had been conceptualized, if not used to its full potential, as early as 1268 by Roger Bacon, it took a Dutch optician by the name of Hans Lippershey in the sixteenth century to assemble lenses in the form of a telescope. Galileo improved upon this early telescope, but as an aid for astronomy—to make massive but distant objects appear closer.

At the end of the sixteenth century, in England, the first microscope was accidentally discovered when a telescope was inverted and it became apparent that it made tiny but nearby objects appear bigger. Modifications soon followed, including the addition of further lenses, principally by Zaccharias Janssen and his father, Hans. The seventeenth century was primed for microscopic study.

The two most important papers on microscopy were published in 1660 and 1665, by the Italian physiologist Marcello Malpighi and English scientist Robert Hooke, respectively. Malpighi set the course for physiology, describing the different tissue types and internal organizations in animals, which will become important later in this chapter. Hooke, meanwhile, discovered the cell while observing a sliver of cork.

Perhaps the most famous enthusiast of the early microscope was Antonie van Leeuwenhoek, a Dutch lens maker with no scientific background. Leeuwenhoek devoted fifty years to microscopy, grinding and polishing his own lenses to reach a new level of perfection. He achieved magnifications between ×30 and ×270, with a resolving power of up to 1.4 micrometers—a fiftieth of a human hair's width—using a secret technique for lighting and viewing his subjects. And so his discoveries began. He published in unprecedented detail images of objects from the animal, vegetable, and mineral worlds, cutting thin sections with a razor. His first publication revealed the structure of mold, as well as the bee and the louse. Then, a few years later, in a measure of water as small as a single drop, Leeuwenhoek unexpectedly saw signs of life. Describing them simply as "little animals," he had observed bacteria and protozoa, laying the foundations for the sciences of bacteriology and protozoology. Leeuwenhoek also found "little animals" in other bodies of water, including lakes, and on the surface of pepper and teeth. He, more than anyone, unveiled the microscopic world. In his bacteria, he had observed the living relatives of those first, single-celled life forms—the earliest life on earth.

The microscope was certainly essential to understanding single-celled beings such as bacteria, and more recently the electron microscope has revealed even more to us in this field. But the light microscope was just as important for understanding more complex, multicelled animals and plants: We still need to understand their individual cells, what goes on inside them and how they are organized. Light and our eyes were the basic rigging needed to know of and to understand life's most primitive forms. Light just keeps cropping up when we ponder life's history—from how we came to know about it all, to how light radiation was crucial to the beginning of everything, to how sunlight shaped the whole tree of life. To understand the next key stage involving light we must look at how single-celled organisms evolved into multicelled animals.

In the latter half of the twentieth century, the field of cell biology revealed how DNA is translated into proteins within the cell. The machinery of life arises from interactions between DNA (or RNA) and protein molecules, whose functions depend on the three-dimensional shapes into which they fold. The larger the DNA, the more proteins can be made and the more functions the cells can perform. Indeed, evolution leaves a trail of animals increasing in complexity. But how and why did those single-celled forms join together to form multicelled animals, and then evolve further into different species?

A species was first defined as a group of like individuals that are sufficiently different from other individuals to be recognizable as such. This definition is difficult to apply when closely related species look very alike, particularly where there exists variation within each species. So a new definition was devised: A species is a taxonomic group whose members can and will interbreed in their natural environment. Even this is useful only for those species that sexually reproduce. Anyway, "species" describes the basic level of categorization in biology.

Independent cells—belonging to a single-celled species—did at some point join together and form a multicelled body. The advantages of doing this must have outweighed the disadvantages. Presumably, a sphere of cells, being a larger structure, would have been less vulnerable to the harshness of the environment, such as the churning of water near the shore. Or maybe it could tolerate a malfunction in an individual cell—its ability to beat the hairs on its outer surface, for instance—where neighboring cells could continue that function without it. Certainly, the genes must have been extremely similar within each cell. Only genetically near-identical cells with a common mutation that favors coalescence would have been involved. Anyway, join together they did, and today there exist reminders of this ancient stage in life's history, such as the species *Volvox*—a green algae consisting essentially of joined-up cells that form the surface of a hollow sphere—in which each cell in the colony is identical and can live independently if separated from the pack.

As the colony divided mutations must have occurred, to the point where, in one generation, the cells were no longer independent. Now the cells needed each other to survive. A division of labor became established so that each individual cell in the colony performed a different job. The first multicelled organisms had evolved. The next stage was the appearance of tissues—groups of identical cells all doing the same job—with two tissue layers in the individual. And at the next evolutionary stage there were three.

Real evidence of the first multicelled animals does not appear in the fossil record until around 635 million years ago. Their evolution was clearly an unlikely event, since it took some 1,000 million years to pass from the stage of *Volvox*-like cell colonies to these first truly multicelled forms, with different cell types. It was the evolution of the DNA to control different cell types that made the odds so long, although the evolution of the ability in cells to communicate to each other was just as improbable. But, given enough time and space, even the most improbable things can happen. In this case, the key was the evolution of controls over gene *expression*—tool kits that activate genes at different times during an individual's development from one cell to many.

Early multicelled animals eventually evolved wonderfully efficient systems in their body for feeding and for sending messages to different parts of the body: That is, they evolved gut and nervous systems. They were now equipped to compete for space on the earth's surface with the microbes—the single-celled life forms that had dominated the planet for some 3,000 million years. They would fill the earth's surface with elegantly complex food webs and audacious lifestyles, as evolution reached a new level of dynamism.

I mentioned that some animal fossils were 635 million years old. How can we ascertain such a precise age so accurately? It is important to establish that there is a robust method behind these calculations, since science bulldozes the foundations of the "young earth" creationists on this point. The answer lies in a discovery of the early twentieth century, radiometric dating.

Certain naturally occurring elements are radioactive, and decay at predictable rates. That is, the atoms of the "parent" element lose particles to become stable "daughter" elements gradually over geological time. Chemists measure the *half-life* of these elements—how long it takes for half of the parent atoms in a sample to decay into the daughter element. By comparing the relative proportions of parent and daughter elements in a rock, its age can be calculated. Although the famous element "carbon-14" can be used for dating rocks only as old as 70,000 years, other elements such as argon (or a series of elements) can give us reliable dates extending back thousands of millions of years.

So, then, what were those archaic multicelled animals at this point? Slugs. A bunch of slug-, jellyfish-, worm- or anemone-like forms—without skeletons, shells, or scales. We first learned about these early, soft-bodied animals through the "Ediacara" discovery.

Sir Douglas Mawson was one of Australia's great explorers and geologists, working from the University of Adelaide. In the first half of the twentieth century Mawson was famous for his expeditions to Antarctica, but he had also found fossil cyanobacteria in the Flinders Ranges near Adelaide in South Australia, and helped to give a date to life's earliest stages. He was also fiercely critical of claims that indistinct fossils were something quite definite, a habit he particularly deprecated in his students.

One of those young students, Reginald C. "Reg" Sprigg, retraced his mentor's footsteps in the Flinders Ranges, looking for further signs of early life. Sprigg found what he thought were the parts of arthropods—the group to which crabs and flies belong—in rocks as much as 570 million years old. But the fossils were not absolutely clear and, as expected, Mawson disapproved of his attempt to identify them. The fossils were bagged up and placed next to the trash can. At the time, scientists believed it was impossible for life forms of this period, with their soft bodies, to be preserved in the ancient rocks.

Undeterred, Sprigg continued to work in the Ediacara Hills, west of the Flinders Ranges. As a young government geologist attached to the newly formed Geological Survey of South Australia, his job was to reevaluate the mineral field; but secretly he was searching for fossils, unable to wrench his mind from the quest for more signs of early life on earth. In 1946, at an old minefield in the Ediacara Hills, Sprigg made one of the most important fossil discoveries of the century. As he traversed a plain sprinkled with weathered sandstone, he noticed strange shapes in the rock surfaces. His initial feeling was that these were the fossils of jellyfish.

Later that year, Sprigg presented his "fossil jellyfish" at an Australian and New Zealand science conference, but was met only with opposition and dismay. Mawson attended the conference, and insisted that these were pseudofossils—marks made by the weathering of rock or by chemical changes that had taken place over time. Others at the gathering reacted similarly, possibly following the influential Mawson's lead. But Sprigg remained resolute. He began examining other fossils accepted as jellyfish from other periods and localities, including some that were 180 million years old. He found many similarities with his own from the Ediacara Hills—too many to be a coincidence. It seemed he had proved his academic superiors wrong.

I have examined Sprigg's "jellyfish fossils" for myself and must admit that at first sight it seems Mawson had a point. They really are nothing but concentric circles with some strange squiggles and folds. To make matters worse, when one places them under a microscope to examine the finer details, one sees . . . a pile of sand. The coarse mineral grain in this sandstone means that nothing finer than that seen by the naked eye has been preserved. Then again, recognized jellyfish fossils can indeed be preserved in this way, as Sprigg had noted from his comparisons. Jellyfish are most graceful as they drift through the open water, using rhythmic pulsations of their delicate bodies. But walk around a tropical island and notice their sun-baked bodies where they have been washed ashore—they look just like some of the Ediacaran fossils.

Of the thirty-seven major groups of multicelled animals on the planet known as phyla (singular "phylum"), the jellyfish belong to one of the first groups to evolve. The jellyfish are the first animals to evolve with three layers of tissues—only the two-tissue-layered sponges are more primitive. Those jellyfish alive today are rhythmical reminders of our ancient beginnings. But there may be other animal groups represented in the Ediacaran fauna too, animals that we can also infer should have existed at that moment in geological time.

Wild kangaroos and dingoes could not have dragged Sprigg from further forays into the Ediacara Hills. His new rock samples—and even old ones, rescued from the bags destined for disposal in Mawson's lab—revealed signs of other life forms.

In his field notebook, next to a sketch of one of the new fossils,

Sprigg recorded: "Ben Dickinson [then head of the South Australian Geological Survey] thrilled with the specimen . . . certainly no doubt whatsoever of its organic origin." The term "organic" is used to contradict the assertion, made by Mawson and others, that these marks in the rock were made later by other mechanisms and are irrelevant to the search for signs of early life. Sprigg also described this new fossil as a "creeper"—a form that moved by its "own power." The animal referred to here became known as *Dickinsonia*. It has since turned up in other localities around the world of similar age, and has become something of an iconic representative of this fauna. *Dickinsonia* looks like a flatworm, but with a division through its center, indicative of the "bilatarian" animals—groups symmetrical down the middle, with a front end and a back end, as well as a topside and underside. Jellyfish, for instance, are radially symmetrical, so that, although they too have a topside and an underside, they lack a front and a back. *Dickinsonia* comes with further lines that represent divisions along the length of its body. The living animal would have resembled a miniature quilted mattress, the size of a skimming stone, itself skimming through the soft sediment of the seafloor. Indeed, even the trails it left behind have been preserved among the Ediacaran fossils.

Sprigg also dug out many "flying saucer" forms. Some would have floated in the water, and these probably represented more jellyfish. But others would have been fixed to the seafloor and were more likely anemones—relatives of jellyfish, using their ring of small, waving tentacles to attract food particles toward their mouths.

Then there were meter-long, leaflike animals that closely resembled the sea pens of today—also relatives of jellyfish, anemones, and corals—and a variety of "worms." The "worms" included *Dickinsonia*, *Kimberella*, and *Spriggina*. Of these, *Dickinsonia* may have been a polychaete worm (related to the parasitic types of today)—a member of the "annelid" phylum to which lugworms and earthworms belong. *Kimberella* may have been a soft-shelled "mollusk"—the phylum to which snails belong. *Spriggina* may have been an "arthropod"—the phylum that contains today's shrimps, horseshoe crabs, and insects. The point is that many of the different phyla of animals had already evolved and were alive in the Ediacaran period, some 570 million years ago, but they all appeared fairly similar. Animals that appear quite different today—lugworms, snails, and horseshoe crabs—actually looked quite similar in the Ediacaran. They had diversified considerably inside, but little on the outside. They were all soft-bodied.

So what had Sprigg really achieved? Because he had found a complete primeval seafloor, he had been able to reveal a snapshot of a community of animals that lived together—roaming the same territory, covering one another's trails. This was far more valuable than trying to piece together a picture of life using individual fossils from different localities and *roughly* the same period. No, these Ediacaran animals all lived at exactly the same moment in time, around 570 million years ago, in a warm tropical sea that covered what is now the Flinders Ranges. Some of these animals became trapped in fine silt in tidal flats and were fossilized as the silt turned to stone. Later, plate tectonics lifted the ancient seafloor into the air. And there was something else especially important about what Sprigg had found. Before the Ediacaran fossils were known, it was thought that only animals with hard, skeletal parts could be preserved as fossils. So the Ediacaran discovery gave scientists a new understanding of the evolution of life on earth, a view of the forms life had taken *before* those hard skeletons, scattered higher up in the fossil column, had evolved. But this new insight brought some controversy in its wake.

Since their discovery by Reg Sprigg, the Ediacaran fossils have been the subject of an evolutionary dispute among paleontologists. The question is whether the Ediacaran fauna are ancestors of today's multicelled animals or whether they represent a "failed experiment" by nature, a line of development that died out. Certainly, as I have mentioned, some resembled modern jellyfish, soft corals, lugworms, snails, and horseshoe crabs, but others bore no resemblance to any living animal group and remain unclassified. As work on classification continues today, the postulated affinities of many of the Ediacaran animals continue to change. For instance, in 1970 Sprigg believed that the fossil *Ediacaria flindersi*, probably the commonest and most widespread of all the Ediacaran fossils, was formed by a creature resembling a jellyfish. It is now commonly believed that the disk shape was formed by a circular "foot" that secured a polyp or soft coral to the seafloor. Nevertheless, jellyfish and corals belong to the same phylum, so this case is neither here nor there.

In 2004 the Ediacara Hills gave their name to a new era of geological time—the Ediacaran era—and were identified for all to see by a large spike in the ground. "Ediacara," the local Aboriginal word for a spring of water that gave the hills their name, has entered the history of science. The fragile fossils themselves are now housed in the South Australian Museum in Adelaide.

The Ediacaran period spans an interval of around 88 million years, from 630 to 542 million years ago. The start of this period corresponds to the end of a worldwide glaciation known as "Snowball Earth" and was accompanied by significant changes in carbon levels. The rapid increase in abundance, size, complexity, and diversity of life forms during this time shows that the earth underwent a period of major evolutionary change. This may be no coincidence: As the ice thawed, the seas became conducive to life. Maybe *that* was the reason why animals remained single-celled for 1,000 million years.

New fossils of Ediacaran organisms have now been discovered in thirty or so localities over five continents, including England, Namibia, China, India, Iran, and Russia, though the principal occurrence remains that in South Australia's Ediacara Hills. Ironically, some of the earliest multicelled animal fossils, dating from 575 to 560 million years ago, have been discovered in Newfoundland, Henry Gosse's early hunting ground in Canada. Gosse spent much of his career arguing that the animals living on the land above these ancient fossils, along with the fossils themselves, were all created by God some six thousand years ago. Alas, modern scientific dating methods cannot reasonably be doubted.

So now we have seen the evolution of the oldest, "plantlike" organisms into more complicated multicelled creatures. But how does this answer the question of lights raised by the Genesis account? Before we can answer that, we need to unlock some of the mysteries of DNA.

Seven years after Sprigg famously applied his geological hammer in South Australia, on the other side of the globe a discovery of perhaps greater significance to humankind took place. It was on February 28, 1953, that Francis Crick walked into the Eagle pub in Cambridge and, speaking for himself and James Watson, announced that "we had found the secret of life." That morning, Watson and Crick had worked out the structure of DNA. But this secret of life had little to do with that revealed by Sprigg—in fact, neither of the two parties would have known what the other was talking about. Crick and Watson had found that the "double helix" could "unzip" to make copies of itself, confirming suspicions that DNA carried life's hereditary information. Yet it was not until decades later, in the age of genetic engineering, that the power unleashed in their lab would be put to world-shattering use.

Many life-saving applications followed the discovery of how

DNA worked. Others, though they attracted less popular attention, were equally exciting to scientists. One of these emerged in the 1980s. This was the concept of a molecular clock, which made direct use of that hereditary information tied up in today's genes to work out when animal groups first evolved and diverged from each other. Each animal species can be distinguished by a standardized DNA sequence, so much so that today we are "barcoding" life by recording these DNA sequences. Using the "bar codes" of modern species from different branches of the evolutionary tree, and assuming that genes (the lines in the bar code) mutate or alter at a constant rate, it is possible by subtracting one line of the bar code at a time to wind back the evolutionary clock and travel through geological time. And the result, according to the molecular clock? The basic types, or phyla, of animals evolved gradually over tens or hundreds of millions of years, somewhere between 700 million and 550 million years ago.

Certainly, by the time the Ediacaran animals lived, 570 million years ago, the seas would have been graced by jellyfish, if not by most of the animal phyla. And even before jellyfish, the most primitive of all the animal phyla would have evolved: the sponges.

Sponges have just two layers of tissues, and they do not move: They suck in seawater with food particles, and spit it out without the particles. And that's it, if you're a sponge. But then another tissue layer evolved to form the jellyfish/coral/anemone phylum and the "comb-jelly" phylum. Then a change took place in one of those tissue layers—a nervous system evolved—and suddenly there is another phylum, a worm of some kind. And so, with further key changes taking place in the tissue layers, the evolutionary tree continues to diverge, until it has thirty-seven branches, representing the phyla.

The key point about a phylum is that it is distinguished by its members' innards, rather than their appearance from the outside as we see them. Now it is clear why the microscope was such an important invention for this subject. The members of each phylum share the same internal organization of tissue types, organs, sensory complexity, and digestive and blood systems—the "body plan" that is exclusive to them. At this level, we should not judge an animal from its appearance—this is not a good indication of its internal body plan or phylum. But it's the body plan that matters when we think of those principal branches on the tree of life.

Of course, once evolved, the principal branches split further. The "chordate" phylum, to which we belong, divided into fishes, amphibians, reptiles, birds, and mammals. Then each of these ramified further—the reptiles into lizards, snakes, and turtles . . . and so on. But as far as we are concerned in this book, it's the phyla only that matter—all thirty-seven of them.

Many of us may think we have a broad zoological knowledge in that we could name many species of animals. But almost certainly this list would contain only the household names—the common exhibits in a zoo, for instance, which all belong to just five or six of the thirty-seven phyla of animals. The remaining thirty or so phyla would appear alien to most of us, yet they form most of the branches on the tree of animal life.

There are moss animals, water bears, velvet worms, hairworms, ribbon worms, peanut worms, acorn worms, arrowworms and flatworms, for instance, each of which have their own phylum. These are just some of the phyla that don't make the headlines and aren't even vaguely familiar to most nonbiologists. An exception could be the flatworms, given that tapeworms occasionally break into the news.

The only phyla most people are really familiar with are the "chordates," to which we belong; the "arthropods," including the insects, spiders, and crabs; the "mollusks," including snails and squid; the "echinoderms," including starfish (sea stars) and sea urchins; and the sponges and jellyfish, which we have been discussing. Most zoos would be filled with these animals only. *These* are what the non-biologist would invariably mean by "animals"—one-fifth of the whole story.

We now know that by 550 million years ago all thirty-seven phyla of animals had evolved, but that they all possessed the body shape of a worm—a worm is a convenient shape—regardless of their differences on the inside. They were also all marine. A modern-day zoo would have seemed a long way off.

It was the great French biologist and paleontologist Georges Cuvier who, at the end of the eighteenth century, began to make sense of the zoological ark. Cuvier was aided by the naming system constructed by the Swedish biologist Carl Linnaeus, in which every living individual was given two Latin or Latinized names—one for the genus and one for the species. Closely related species would share the same genus. At last we could see what was out there in systematic form, rather than gazing helplessly at a mass of individuals in a very tangled web. Linnaeus had cleared the clouds—but Buffon and Lamarck had introduced a red herring: That the structure of animals was changeable, affected by their environment. Cuvier had the foresight to discount this hypothesis.

Cuvier began his career to the opening gunshots of the French Revolution. He took refuge as a tutor in Normandy to avoid the worst of the violence, and soon joined the local government there and developed a reputation as a naturalist. At the end of the eighteenth century he was invited to the newly formed National Museum of Natural History in Paris as an assistant in animal anatomy, and quickly became a professor. When Napoleon came to power, Cuvier received several promotions to high positions in the government, and continued as a state councillor under three successive kings of France.

Working away in his lab, Cuvier discovered that four groups—chordates, arthropods, mollusks, and jellyfish—each possessed their own, unique internal organization. This finding was of immense importance. The concept of the phylum—the highest classification of animals, at the other end of the hierarchy from the species—was born, and biology accelerated from there. One by one, the remaining phyla were recognized. It was now, given the classification theory of the phylum, all so simple.

Returning to the later evolutionary timeline of animals, thanks to Sprigg's discovery, clarity was emerging there too. Now we knew that animals had evolved gradually over 450 million years to the point where all of the phyla were in place, but all still had the soft body of a worm. A worm-shaped body is a convenient form, and hard, skeletal parts are energy-expensive in comparison.

THE FIRST EYE—AND WHY IT MATTERED

Vision mattered very much in the evolution of life. It is ultimately the answer to the question about those mysterious "lights" in Genesis 1. So what do ancient "worms" have to do with eyes? Well, one of those worms, a rather flattened variety, known as a proto-trilobite—possibly even *Spriggina* or a relative thereof—began to evolve an eye.

In a special section entitled "Organs of Extreme Perfection and Complication," Charles Darwin wrote in *On the Origin of Species* in 1859: "To suppose that the eye . . . could have been formed by natural selection seems, I freely confess, absurd in the highest possible degree." Let's consider what an eye really is.

Slotting a lens into the body of a digital SLR camera, complete with its light-sensing array, and connecting the camera to a computer for image processing, one is reminded of the human visual system. The human eye and associated visual cortex in the brain mirror this camera in its lens, retina, and processing unit. But that's the crux of Darwin's dilemma—the visual system comes in three unrelated parts. For such an eye to evolve, all three parts must evolve independently and simultaneously. That, at first, does appear absurd. There are two reasons, however, why it is not. The first involves the way the eye evolved; the second concerns Darwin's thoughts on "perfection."

Eyes detect the peak in the sun's output of electromagnetic radiation—light—and come with limited architectures because of constraints set by the physics of light. They are the sensory organs at the beginning of visual systems.

A visual system such as our own, with its associated processing unit—the visual cortex in the brain—is a complex of parts that have very different functions and are unrelated in terms of their initial evolutionary history. A lens focuses light to form an image, retinal cells convert light waves to electrical signals, optic nerves carry those signals to the brain, one part of the brain reconstructs the image and constructs color, and another part of the brain makes a decision on how the body will react to that image.

Eyes come in either chamber or compound types. Chamber eyes, like our own, have a single entrance for light—usually with a single, large lens in front of a concave retina. Compound eyes, like those of a fly, have multiple entrances—many tiny lenses, in some cases seeing their own small portions of the world that are combined like pieces of a jigsaw puzzle to form a complete image. But both chamber and compound types produce an image on a retina—light sensors do not, and so cannot be regarded as eyes, as I'll explain below. In fact, the evolution of simple light sensors to become the first eyes during geological time marks an important transition for life.

The question "When did the very first eye evolve?" has traditionally not been posed, let alone answered. Yet that question refers to the moment when vision was introduced on earth—and that is important.

For the first few hundred million years of multicelled life on earth, animals, including those of the Ediacaran period, were blind, yet the "selection pressures" for vision were always immense. Animals could always theoretically benefit from sight, although of course evolution does not strive to produce the best; it has no inherent purpose at all. If a mutation that occurs during DNA replication leads to a lens, then, if the host animal can gain enough from that structure to outweigh the high energy costs of building and maintaining an eye, that mutation will begin to spread through the population over subsequent generations.

That very first eye evolved in a soft-bodied trilobite. This may have been a relative of *Spriggina*, or a similar form discovered in southern Morocco of about the same age, but it was certainly a distant relative of today's horseshoe crabs. At around 521 million years ago, this soft, flattened animal formed crystals of calcite in its skin. The elements in calcite were probably waste products that became trapped within the skin cells. In the head region of the proto-trilobite there would have already been simple light sensors, which exist in even single-celled forms, along with soft-bodied worms and in fact most animals without eyes. But now the calcite crystals were focusing light onto the proto-trilobite's light sensors more intensely. Suddenly any mutation that led to improved focusing or processing of the visual information would also have spread within the population over generations. With enough mutations, as I will explain, the proto-trilobite evolved an eye.

Prior to this, most animals—not just the trilobites—would have possessed only simple light sensors. This is where the difference between an eye and a light sensor really matters. Light sensors provide information on night and day, or up and down in the water. Think about how, with our eyes lightly closed, we can still slowly locate a light in a room. Eyes, on the other hand, allow animals to interact with each other using light as a stimulus. With eyes, an animal can instantly distinguish between friend and foe. And then it can track the foe if it would like to eat it.

Eyes have evolved independently many times since, but that first eye to evolve—in the soft-bodied trilobite—let loose the sense of vision on earth. Suddenly "light" meant something altogether different. Unlike the other senses, which evolved gradually through geological time, sight appeared with a bang. At the moment a lens evolved, an image was formed—that animal had the potential to see, and would soon have evolved the ability to use that image. The consequences were huge.

The argument that eyes cannot evolve because the intermediate stages in organ evolution could not function does not hold up. At the University of Lund in Sweden, a computer model was made by Dan-Eric Nilsson and Susanne Pelger to show how a patch of light-sensitive skin could gradually transform into a chamber eye within—in a pessimistic estimate—2,000 sequential changes, or just 364,000 generations. Certainly, there have been enough individuals within a species and geological time to accommodate this. With the foot down on the pedal for eye selection, the organ alters first to distinguish between sunlight-on and sunlight-off, then to sense the direction of sunlight, and finally to form an increasingly well-resolved image. The important point is that each stage confers an improvement in light perception or vision over its predecessor.

It is only in the very final phase of focus tuning that the spatial resolution—the number of pixels formed—shoots to extremely high values with only minor changes in the eye's structure. The evolution of a lens would have at some point provided a "leap" in the information supplied by the eye, but animals don't constantly use all the information their eyes deliver. Human vision fails to make full use of the retinal image over most of its field of view. It is plausible that the evolution of the central nervous system's processing of visual information lagged behind the available information for a lot of evolutionary time during the evolution of vision.

Recent research based on advances in molecular biology has centered on photoreceptor evolution. It was found that the gene that encodes opsin—the protein component of the light-sensitive visual pigment—and some developmental genes date back beyond the first eye and were recruited repeatedly during the evolution of different eyes, possibly between forty and sixty-five times. Because the same development tool kit was used each time (and this can be found within all animals anyway), it is no problem to explain why the eye evolved so many times. We can demonstrate the role of these developmental genes in the fruit fly *Drosophila* by experimentally activating them in different cells within a developing embryo, and find eyes sprouting on legs and wings in the adult.

The evolution of lenses may owe something to physical phenomena such as crystallization or similar sorts of molecular self-assembly. If an enzyme, such as a liver enzyme, is overproduced within a cell it will crystallize—our own lenses are formed this way. Calcitic lenses in brittlestars—relatives of starfish, or sea stars—also evolved in this manner, and those in the very first eyes in the proto-trilobites very probably did likewise.

This is all extremely relevant to the science versus religion debate. Let us return to that statement by Darwin about the eye as an exceptionally complex and perfect organ. In particular, we should focus on his use of the word "perfection" when referring to the eye's optical achievements.

How the different parts to the visual system came together *perfectly* may be difficult to explain, but an *imperfect* system is more easily elucidated. We touched upon this point above in mentioning the underutilization of information available—and indeed, there are signs of imperfections in all eyes. To take just one more example: The human retina is inside out, with nerve fibers running back into the path of light rather than behind the retina—as they do, more logically, in the similarly structured eyes of squid.

Imperfections can also include failure to deliver on visual requirements. Such shortcomings can be revealed within the "arms race" that exists between the eye of a predator and the color of its prey, for instance. We have trichromatic vision—three types of cones in the retina, each responding to a different wavelength range—and as a result we can see cuttlefish resting on the seafloor. But their predators cannot. Those predators are dolphins with monochromatic vision—they see in "black and white," under which circumstances the cuttlefish are perfectly camouflaged. Another limitation of the *human* visual system is that we do not see ultraviolet light, and consequently miss out on the ultraviolet patterns of flowers that attract insects, and the ultraviolet patterns of insects that attract birds. The point is that the words "extreme perfection" used by Darwin are inappropriate with respect to the eye. He would have been troubled less by the eye if he had understood that no eye is ideal.

But Darwin would have also been unaware of most of the fine details outlined in this chapter so far. He would have visualized how single-celled forms lay at the base of the tree of life, considering their obvious simplicity. But he would not have known that the earliest cells were plantlike, nor how individual cells later joined together to form a multicelled animal, nor that trilobites were the first animals to evolve eyes, after a period where the evolutionary tree took on its main branches but the animals remained soft-bodied. If he had known this, he would also have been untroubled by the lack of early fossils, since soft-bodied animals generally do not leave fossils. And it's only when we know all this that we find a strong parallel between the scientific account and Genesis 1 . . . at least so far.

But to return to the eye, and its failure to gather all the visual contents of an environment: What, then, of the information the eye *does* collect from its surroundings? Descartes had something to say about that.

Descartes, introduced in chapter 4, made the statement "Cogito, ergo sum"—"I think therefore I am," or "I am thinking therefore I exist"—to summarize his ideas that thought is something that exists, and that the human body cannot be separated from thought. He concluded that he could be certain he existed because he thought. But his questioning did not stop there. He next asked: In what form did he exist? His answer was extremely relevant to the Genesis Enigma.

Descartes understood that he perceived his body through his senses, and his senses only. But the senses were already known to be unreliable. The eye is not only limited as to what it sees, it even distorts the information it *does* collect from the environment. Waves of electromagnetic radiation become colors, for instance (but more about that later). Given that he could not trust his senses—a difficult concept to grasp—Descartes understood that the only thing about him that could not be doubted was his thinking.

He demonstrated this idea by using wax as his model. Wax is usually hard but pliable, and has a certain texture, smell, and depth of color. This is what our senses tell us. Yet when the wax is heated, many of those characteristics change—our senses tell us something else, yet the material is the same. Still, we know it is wax whether molten or solid. What is telling us this? The mind.

Fortunately, Descartes assured us that the world around us, although misconceived, actually does exist. He reasoned that the sensory stimuli—light, sound, smells—came to him involuntarily, and were not willed by him. They were external to his senses, and so something did exist outside his mind and body—an external world. Still, the world does not exist as we think it does.

We take for granted that everything around us is just as it seems—green grass and solid rocks. This is a feeling in itself. But it is not true. Our senses are misleading us. They collect *some* of the information available in our surroundings, such as some portion of light, and convert it, in the brain, into a form in which we can make best use of it through our limited processing capacity, or brain space. But we can never be aware of this. We can never understand that we only know a portion of what's really there, and that even this is distorted into a form unrecognizable from the truth. We can never properly take this into account.

Credit is due also to Henry Gosse who, in 1857, made the same inference in the opening to *Omphalos*. Gosse pointed out that to know anything about our world we are completely dependent upon our senses, and we are forced to fully accept what they tell us, deficiencies and all. Unfortunately Gosse continued this line of reasoning until it became a teleological argument. He used the limitations of the senses to back up his belief that God had created every living and fossil species—Gosse had made up his mind. But if he had stuck to the concept of limited senses, and had remained open-minded, he would have found in this notion some reason to consider the existence of something beyond that which we deduce from science, albeit not necessarily God (and Gosse, to his discredit, was only interested in God). That there may be something in the universe—something that we do not detect, and that is not revealed through scientific treatment—is a possibility, but is not necessarily evidence of God. I will save closer examination of this point for the final chapter of this book.

For now, let's link T. E. Lawrence's comment that "machines are extensions of our faculties" with my earlier statement that "the world we experience and know does not really exist." In the history of life, some 521 million years ago, an eye evolved in an animal. This was *a natural extension of its faculties*, but a colossal one; remember too that *of the five senses, the most important is sight*. Suddenly this species had sight. This presented both a potentially supreme advantage—and yet a false view of its surroundings. Now, more than ever, an animal would live in a world of virtual reality—virtual because what it thinks is there, is actually not. The dolphin thought only the seafloor stood before it, where really there also lay a cuttlefish. We think we see everything there is to see, yet miss out on the world of ultraviolet. Rocks appear completely solid but are in fact 99.9 percent nothing (because of the empty space within atoms, as explained above). Leaves appear green, regardless of the fact that there's no such thing as color in the real world. But that's the way it happened, and virtual reality would become a thing of the future.

In the paragraph above I used the term "potentially" as a prefix to the "supreme advantage" that sight would have bestowed on its host. The advantage was only a potential one because the animal that evolved those eyes was a flat, listless, soft-bodied form that spent its days writhing across the seafloor. What use was an eye to this creature? Well, it would have been useful in pinpointing its food, which lay scattered round about it. But it would also have seen, for the very first time, all of its neighbors. It would have seen all the other sluggish, soft-bodied animals on the seafloor around it. And that's the "potential." These other animals could also be viewed as defenseless chunks of protein, just waiting to be eaten.

Evolutionary language speaks of "selection pressures"—invisible forces in the environment to which natural selection reacts. If a potential does exist, then one of the myriad mutations, constantly stirring at each generational turnover, will surely realize it, sooner or later. The selection pressure acting on the soft-bodied trilobite was to become a predator. And, sure enough, the proto-trilobites with eyes did indeed evolve into efficient predators—they gave way to the trilobites.

The same mechanism of holding on to their calcitic waste products to form the lenses in their eyes—a process known as biomineralization—began to work its magic on the trilobites' entire bodies. The trilobites became fully armored, and their hard-shelled bodies were further endowed with oars—they became streamlined and equipped for rapid swimming. Since no other animal could move with any great speed, problems clearly lay not far ahead. To make matters worse, the trilobites' legs resembled those of a suit of armor . . . with sharp endings. More about those sharp tools in the next chapter.

The parallel between science and scripture in this chapter is staggering. The introduction of day and night translates, in biological terms, to the evolution of vision, which scientists considered a monumental event only as late as 1998. The very first eye on earth effectively turned on the lights for animal behavior and consequently for further rapid evolution, while providing accurate recognition of night and day. Prior to the eye, the light that existed, radiating from the sun, had no effect on animal behavior. This was the country—the planet—of the blind. As a result, animal bodies remained simple, mainly wormlike, and interactions between animals were almost nonexistent. The whole system or ecology of life on earth was uncomplicated, vacant of the type of complexity we find today. With the advent of vision, all that was set to change.

I have just mentioned the light radiating from the sun. Since the proto-trilobites and the trilobites were marine animals, to reach their eyes it had to pass through water, too. And here we meet again the coincidence, mentioned in chapter 4, of the peak in the sun's intensity lying in the same region of its output as that which transmits best through water. It turned out that what we call sunlight—actually a very narrow band in the sun's vast range of radiation, but also its most intense band and so potentially the most useful to life—just happens to travel through the sea particularly well. Life—exclusively aquatic at this point—was presented with the most beneficial of the sun's radiation. Evolution capitalized upon this, and eyes detected this very region of radiation—the visible light range. This coincidence was extremely convenient for life, and must take part of the credit for the success of sight, which we will hear more about in the following chapter.

Why the author of the creation account placed emphasis on the sense of vision, and why he chose this moment in his narrative to introduce the eye, cannot be explained. Knowing what this author knew in his day—basically, nothing that I have discussed in this chapter—I certainly would not have opted for this emphasis. Incredibly, though, he remained on a parallel course with the scientifically correct sequence of events in the history of life, and had highlighted life's most dramatic innovation. The reason for this—the consequences of vision—will become evident in the following chapters.

The copy of T. E. Lawrence's classic *The Seven Pillars of Wisdom* that I found in the antiquarian bookshop had been inscribed "W. R. Darwin, July 29th 1935." This was the first trade edition, published in the year of Lawrence's death, and was found in the library of Clouds Hill in Dorset, England—his last home. The book's previous owner was William Robert Darwin, a grandson of Charles Darwin.

Despite considerable effort I have failed to find a connection between T. E. Lawrence and Charles Darwin (not personal, for Darwin died in 1882 and Lawrence was born in 1888), although I have found another link between their writings. Think again about Lawrence's comment quoted at the beginning of this chapter, in which he described machines as an extension of our faculties; then consider that in Darwin's classic *On the Origin of Species* eyes were singled out as a similar extension of an animal's faculties that confers a major evolutionary advantage. Using any other sense but vision, a lion could not chase down a zebra, or a pike pursue a roach through a reed-bed. This lesson is that *highly active* predation—predation that involves rapid movements—is usually not possible without vision. Here we have further justification for that other phrase set out early in this chapter, "Of the five senses, the most important is sight."

And then, again, there was the statement: "The world we experience and know does not really exist." That is all about how vision provides an incomplete view of the world around us, in that the eye and visual processing unit in the brain provide us with information in the form that is most useful to us. These mechanisms collect only some of the information available, and then filter and distort it so our body can react to that information in a way that will best help us to survive.

We have now reached the point, 521 million years ago, when the trilobites evolved. We have covered vast swaths of geological time to get here. Remember, the earth formed some 4,600 million years ago, with the first signs of life at 3,900 million years ago. The first photosynthetic forms—the cyanobacteria—appeared around 3,600 million years ago, while cells did not group together until about 1,600 million years ago, when *Volvox*-like forms emerged. At 635 million years ago the first multicelled animals evolved, and thereafter these went on gradually to evolve further into the thirty-seven phyla we have today. Then, at around 521 million years ago, the trilobites evolved: the first animals to possess hard skeletons and eyes. These were the first of the modern body types to appear—animal bodies as we know them. In other words, the evolutionary history covered so far in this book lasted a long, long time—in fact, for most of the history of the earth. The remaining one-tenth of the earth's history will be covered in the following chapters; and although it does not appear a great amount of time when put this way, it hosted a *lot* of activity.

The history of science in this book opened with a Big Bang of cosmology. The next chapter will cover evolution's Big Bang—the Cambrian explosion.

VARIETY

"Let the waters bring forth abundantly the moving creature that hath life."*

Following the trail of events in the creation account of Genesis 1, it is next claimed that the waters brought forth masses of marine life. It is suggested that a fundamental change took place in the sea—a change that resulted in teeming quantities of living creatures, perhaps like those we see on coral reefs today. This is extremely interesting, since the individual who wrote these words, and Moses who may have first spoken them, lived far from the sea and would have known only of the land mammals, birds, and insects with whom he shared his territory, plus perhaps one or two fishes from the sea. But take a second look—in this fifth phrase or stage in the creation account, the animals of the *waters* are referenced specifically and unmistakably. There's no mention of life on land yet, regardless of the fact that this is the moment where the variety of animal life—all those kinds of "creature"—is established.

*The most recent translation (*Today's New International Version*, 2004) gives: "Let the water teem with living creatures."

Why would the writer record such a thing? It goes against all he knows. In fact, it portrays what he does not know. But is it correct, in that, like the previous statements in Genesis 1, it has a parallel in the actual history of life on earth as exposed through science? Let's take a look at what the modern scientific account tells us.

FIGHTING BACK AGAINST THE TRILOBITES

When those first vision-equipped, torpedolike trilobites appeared on the scene, their soft-bodied neighbors became visible, for the first time, as defenseless chunks of protein—potential meals. And the trilobites set about them. Most of those wormlike or globular species would have been eaten to extinction. But some escaped the trilobite terror, and this chapter will explain how. It was a case of evolution saying, "If you can't beat them, join them."

The time when the first trilobites, with eyes, evolved was around 521 million years ago. This fell within a geological period known as the Cambrian. During Darwin's lifetime several fossils from the Cambrian were excavated, including trilobites, but none from before. So did something happen in this period? This chapter is all about the very foundations of the *variety* of life. It is also about digging.

Trilobites, which finally became extinct around 250 million years ago, are distant relatives of spiders and horseshoe crabs. Some 17,000 species of trilobite have been excavated from different parts of the world. In fact, trilobites rank second only to the dinosaurs in fossil fame. Although they ranged in size from ant to anteater and were equipped for a wide range of lifestyles—including hunting—their fossils are always instantly recognizable. They have been known to man for some time: At Arcy-sur-Cure in France archaeologists found a 15,000-year-old amulet in the form of a drilled trilobite fossil. But the science of trilobites began in the eighteenth century, after earlier ideas that they were flatfish. And science could not ignore a characteristic common among them—eyes. The first trilobites were all equipped not only with strong limbs and paddles for speedy swimming, but also a pair of precision eyes. They could see; and they ruled the world.

This brings me to my final reference to T. E. Lawrence. In 1920 Lawrence published an article in the military journal *Army Quarterly* called "The Evolution of a Revolt." Here he detailed his experiences in the Middle East after hanging up his archaeological tools and joining the British army. This was the adventure that brought Lawrence worldwide fame and provided the story for the biographical film *Lawrence of Arabia*. Of interest to us is the term "evolution" in the title of his article—for it does have relevance to the evolution of life in the Cambrian.

About halfway through World War I, in October 1916, Lawrence embarked on a fact-finding mission to the Hejaz, in the northwest of present-day Saudi Arabia, where Sharif Hussein of its holy city Mecca had rebelled against Turkish imperial rule. Lawrence worked hard to gain the confidence of Hussein and secured a role as British liaison officer in the Arab Revolt—the unified Arab attempt to secure independence from the ruling Ottoman Turks.

Lawrence spoke the Arabs' language in every sense. His job was to prevent the Turks from withdrawing from Medina and transferring to the Palestine front to halt the British advance. But the odds were stacked against Lawrence and the Arabs. The Turkish army was far better equipped than Lawrence's makeshift resistance force. Lawrence went through the military manual, but no tactic since the time of Napoleon seemed appropriate in this case. The Turks held vast fortresses and powerful field guns, and their numerous, well-drilled soldiers fired the latest rifles. Lawrence would have to think on his feet.

Then Lawrence wound his memory back further, to his university days when he had studied, step by step, the military campaigns of Hannibal and Belisarius, Mohammed and the Crusaders—in many ways more primitive in their styles. It struck Lawrence that the Turkish army was indeed extremely competent . . . but only against soldiers who fought according to the latest military techniques. They were trained to fight other modern forces—"an army attacking with banners displayed." But how, he wondered, would they fare against something very different, the likes of which they had never encountered before? Lawrence's hand was partly forced. He took a good look at his troops and made the comment, "In the emergency it occurred to me that perhaps the virtue of irregulars lay in depth, not in face." His soldiers were ill equipped in arms and mind, but there were lots of them. So Lawrence set about a completely novel strategy.

"Suppose we were an influence (as we might be), an idea, a thing invulnerable, intangible, without front or back, drifting about like a gas? Armies were like plants, immobile as a whole, firm-rooted, nourished through long stems to the head. We might be a vapor, blowing where we listed," he wrote in "The Evolution of a Revolt."

And so the Arab soldiers became a human gas. They attacked the Hejaz railway, the Turkish supply line running south from Damascus. They became notorious for appearing at any point along the railway, at any time, in any numbers and using any tactic. They were unpredictable. Following an assault, they would disappear into the sand whence they had come. Everything was organized locally, leaving no messages to be intercepted, and there would be no forewarnings ahead of their appearance or trails left behind them.

This was a totally new weapon. The Turks did not know what had hit them. In effect, they could not see their enemy. They were lost for a defense tactic and were rendered impotent in their own territory. Lawrence extended his revolt northward, with equally good effect.

The point is that, no matter how well adapted an army is, if a weapon is launched that is so novel, so unknown, that it leaves that army oblivious and helpless, then it can be defeated. We have, sadly, witnessed a similar tactic in the twenty-first century. But here, let's think about evolution. Let's think, in a similar vein, about that new trilobite with vision that was thrown among the soft-bodied animals 521 million years ago. That will be the aim of this chapter.

But first, let's bid farewell to Lawrence. He wrote up his extraordinary military adventure into a manuscript, some half a million words long. This would become the literary classic *The Seven Pillars of Wisdom*, but not without some further excitement. The briefcase in which Lawrence kept the manuscript was stolen while he was changing trains at Reading station in England in 1919. Despite pleas in the press, the manuscript was never returned. Lawrence was left to rewrite his masterpiece from memory, since he had destroyed his diaries and notes after completing his first draft. Fortunately he satisfied himself that he had made a near-perfect copy.

Lawrence may have written the article "The Evolution of a Revolt" as a summary of his endeavors in the desert in order to claim them as his own should anyone else attempt to pass the stolen manuscript for *The Seven Pillars of Wisdom* off as their own. Fortunately that never happened, and its whereabouts remain a mystery to this day.

We have taken a diversion to Arabia with Lawrence because his story offers a useful parallel with the moment in life's history we have reached here. Now we want to know how those soft-bodied "Ediacaran" animals evolved in the geological time that followed the advent of *their* new enemy, the trilobites with vision. We got our first glimpse of this outcome in 1909, at a time when science was working with Darwin's and Wallace's idea of microevolution—that evolution took place gradually and evenly through geological time.

THE BURGESS SHALE AND THE CAMBRIAN EXPLOSION

In the summer of 1909 Charles Doolittle Walcott, secretary of the Smithsonian Institution, set off to the Canadian Rockies with his family to look for fossils. Cambrian trilobites had been found in this area, and the Walcotts chose a nearby site just off a new path that had been made to traverse a mountainside. They unloaded their mountain donkey and camped just below the path. Then, as Walcott began splitting open some of the loose, slatelike shale on the mountainside, he saw a strange silhouette emerge, the size of a shirt button.

This was not a trilobite. Walcott did not know what it was, but he could see that its condition was remarkable—even the finest hairs had been preserved. He began splitting more shale, posthaste.

Further weird and wonderful shapes emerged from the grayness of the rock. Walcott had stumbled upon a radically newfangled fossil site 508 million years old—twice as old as the first dinosaurs—which became known as the "Burgess Shale," after Mount Burgess, near where the fossils were found. The animal bodies preserved here were like nothing science had witnessed before.

Walcott and his family continued to discover a whole community of animals. Species were revealed with a variety of lifestyles, from burrowing and running to swimming and parasitizing, and with the most alien body plans one could imagine. *Hallucigenia*, for instance, was so bizarre that researchers would first reconstruct it upside-down and back-to-front, with its sausage-shaped body walking on stiltlike spines and waving a row of tubes into the water above it. Then, when Cambridge University paleontologist Harry Whittington revealed a reconstruction of the Burgess animal *Opabinia* at a scientific conference in the 1970s, it was greeted with loud laughter, such was its Martian visage. Just to begin with, it has five eyes and a long, flexible snout terminating in a jaw.

Eventually, as one might expect, these Burgess fossils would indeed capture the public's imagination—recently they made the cover of *Time* magazine—even though most were smaller than a shrimp. But although they looked like the ultimate space invaders, the bits and pieces that made up their weird and wonderful bodies can still be found, somewhere, in animals today. Unlike the Ediacaran animals of South Australia, from about 570 million years ago, those of the Burgess fossils possessed the characteristics of modern species. The really important point is that the copious number of species preserved in the Burgess Shale represent *many* of today's thirty-seven animal phyla, all preserved together, as in an ancient photograph, in the manner of the Ediacarans. A profusion of different body types had been set in stone, and were now telling us that they did indeed occupy the earth 508 million years ago. Critically, all the Burgess species possessed the hard, skeletal parts that are characteristic of their living relatives. They had evolved from the Ediacarans—the Ediacarans had become armored. But why were these fossils of marine life entombed high up in the Canadian Rockies?

Again, as the earth's plates moved around throughout geological time, and slowly crashed into each other, some things had to give. The rocks that now form the Canadian Rocky Mountains are one example. Originally, they were formed underwater hundreds of millions of years ago by sediment in the sea settling onto the bottom, forming a new seafloor. Following compaction they were forced up from below the water and into the air, turning an ancient environment upside down. But in addition to preserving the sediments of an ancient time, the Rocky Mountains entomb another part of ancient history—the biological one. In the manner of the Ediacaran animals, any body that died on, or fell to, the seafloor would have become covered by the next layer of sediment and potentially preserved as a fossil. But sometimes a particularly heavy layering of the sediment itself can be the cause of death for an animal, in the form of a live burial.

The Burgess fossils were of animals and plants that lived somewhere around 508 million years ago in a well-lit marine reef, at a depth of seventy meters or less. One day in the Cambrian period, an avalanche of very fine mud swept across the area, burying the local reef, including the Burgess fauna and flora. The inhabitants were preserved in all sorts of positions, akin to the bodies at Pompeii when the volcano erupted. The variety of life embraced by these fossils will be put into context below.

The Burgess evidence pointed toward an explosive event in evolution that would shatter Darwin's idea of a gradual evolutionary history. Unaware of the Ediacaran fossils or molecular clock, Walcott concluded that all groups of animals had suddenly evolved in this Cambrian period, whereas in fact only their hard, skeletal parts had evolved; but even this was an explosive and fateful event all the same. No simultaneous development on this scale would ever take place in life's history again.

With hard, skeletal parts in place within each group of animals, lifestyle possibilities were greatly expanded. Rather than simply floating through the water or crawling on the seafloor, thanks to the innovation of hard shells and skeletons, animals could now run over rocks, dart through the water, battle with competitors, ward off predators and grasp victims . . . In other words, animals could do all the things, and interact with other species in all the ways, we find on coral reefs today. As would be expected, the diversity of life exploded. Evolution went into overdrive until every niche had been filled by swarms of species with all manners of body shapes and sizes. *This was evolution's Big Bang—the "Cambrian explosion."* And we owe much of our knowledge of this biological flashpoint to the fossils of the Burgess Shale.

In 1924, the assistant secretary of the Smithsonian Institution made the comment on Walcott's discovery that "This pushes the age of the habitable world upward, quite in step with the recent discoveries in the physics of the radio-active elements, which tell us that for a billion [1,000 million] years, at least, the earth has been full grown." The Ediacaran discovery, of course, lay in the future. But this was the moment when solid evidence, rather than theory, to contradict the biblical seven-day account of creation arrived from different directions, from scientists who knew nothing of each other or each other's subjects. Collectively, Walcott's Burgess fossils—a philosopher's stone—would appear as a nail in God's coffin.

So it was not just the trilobites that evolved hard parts in the Cambrian period, although they do appear to have been the first to do so. Biomineralization had become a trend.

As mentioned in the previous chapter, biomineralization is the process by which animals incorporate minerals into their bodies, leading to the formation of hard, skeletal parts. Our bones, for instance, capture calcium from our food to provide strength. There are two ways in which animals can do this. The first is where the condition of a cell causes a mineral to settle out of its surroundings and onto the cell itself. The second way is "biologically controlled," whereby mineral crystals form within a framework or matrix laid down by the cell. The cell has much more control over this method, and can select specific minerals from its environment, rather than simply grabbing what is most readily available.

Biomineralization was an evolutionary innovation that always lay within the potential of the basic animal cell—even the most primitive sponge cells can extract silica from their environment to make glasslike spines, or "spicules." Simply, biomineralization, or the hard parts that result, was a weapon waiting to happen. It was a little energy-expensive, nonetheless, so without the *need* for hard parts, it remained a latent capability, a process held in reserve, as it were. But when duty called, and selection pressures tipped toward physical protection, genetic mutations that activated the process were suddenly retained—and retained in all phyla. Biomineralization was in a geological instant turned on across the board, in all creatures great and small.

The Burgess fossils possessed hard, skeletal parts like those animals characteristically possess today. The mollusk phylum evolved shells and scales, the starfish and shrimp phyla evolved hard, external skeletons, and the velvet worm and bristle worm phyla evolved sharp spines and shields, for instance. They were all ready to begin their scramble for places on the evolutionary tree, because now life had changed forever. Soft skins were out—the age of the Ediacaran animals had ended. They could no longer compete. Hard, skeletal parts were here to stay.

Whereas the Ediacaran animals, with their totally soft bodies, were extremely fortunate to have left a fossil legacy, once skeletons had evolved in the Cambrian the odds in favor of preservation in rock improved.

So it does appear that the fifth stage in the creation sequence as described in Genesis was precisely correct—that vast numbers of marine creatures appeared. Once again, an event in the history of life matches an event in the creation story. Extraordinarily, the Genesis creation story implicates a *diversity* of sea creatures. There is no way such diversity could have been understood at that time.

Then again, the overlaps between the 508-million-year-old fossils of the Burgess Shale and animals today, and between the hard parts of the different phyla, are indicative of the limited range of processes that take place in animals' bodies. To make hard parts, structures within the cell are called into play. The same structures exist in all multicelled animals, and occasionally a gene will mutate to activate one—to infill it with a mineral, for instance. The point is that animal bodies are not the endless wonder or unfathomable complexity postulated by theologians such as the influential William Paley of Erasmus Darwin's day. They are in fact the result of ancient mechanisms—predetermined in the way the first animal cell came together—occurring and recurring again and again through geological time. Now evolution becomes far simpler to explain. The theory works without the inspiration and intervention of a Creator.

In essence, when the biblical text is taken literally, it is left in the wake of an advancing science. But when it is read figuratively, it not only keeps pace with the hottest science, it precedes or heralds it. I am left to question why theologians should choose to insist upon the literal option, through false logic and shoehorned evidence, when the figurative or symbolic alternative is both honest and supportive of religion. As we have noted previously, human storytellers convert the unfamiliar into meaningful tales—with the exception of the opening page of Genesis, where there may have been no possible alternative to the message as presented.

Ancient texts from biblical times and even earlier, such as ancient Egyptian papyruses, include some descriptions of native wildlife. They highlight the oryx, wild ass, gazelle, ostrich, crocodile, hippopotamus, wild dog, lion, monkey; they mention lizards, snakes, frogs . . . almost the complete zoo that the average person considers as "animals," an analogy already used to place the Ediacaran fossils in perspective. Even the remainder of the Bible mentions few animals beyond these large members of the chordate phylum. So why does Genesis not, as one would surely expect, discuss only the chordates, or at least the five or six phyla that make up the modern zoo, when describing the creation of animals on earth? The answer could be that its writer was recording the true, scientifically accurate story, and not the version understood by the people of his time.

Today, neither scientists nor theists have noticed this fact. I dug out a clipping from the London *Times* published in 1963 titled "Israel reserve for Biblical animals," which reported a new zoo in Israel for wildlife known to Moses and later Hebrew prophets. The article covered the usual suspects—oryx, wild ass, gazelle, and so on—which were to be brought back to Israel, having become extinct in that region. The zoo did not happen in the end, although an alternative did, a Judaic Park in San Diego. Neither the original idea in Israel nor the Californian alternative had plans for the swarms of marine life mentioned in Genesis. Why not? Because these marine animals don't make the news and they aren't household names; so they really don't come to mind at all.

To summarize the point we have got to in the evolutionary story line: At some time in the Cambrian period the animal kingdom burst into life. This was the time of the Cambrian explosion. Prior to this, animals were all soft-bodied, like worms, as they had been for 100 million years since the first multicelled organisms. Then, in a relatively short space on the geological timeline, nearly all the animal phyla on earth today independently evolved hard parts. At no other time in the earth's history has there been such a profusion, such an exuberance, such an overwhelming diversity achieved in so short a time, within about one million years. The big question remaining is: Why? What lit the Cambrian fuse?

For some years now the paleontological and biological communities have been suggesting explanations. These include the ideas that 521 million years ago there happened a dramatic increase in oxygen levels, or just as spectacular a decrease in carbon dioxide levels. Either of these would have been encouraging for life. But neither was the case, as we have learned from a new, accurate dating system. The same problem of poor timing undermined other ideas, such as surges in the availability of the minerals that fueled biomineralization. These can no longer be timed to coincide with the Cambrian explosion.

Another idea, which appeared convincing in its time, was that it was all in the genes—the evolution of the major developmental tool kits, to be precise. But this, like many other suggestions made, was flawed by association with a major red herring—the wrong definition of the Cambrian explosion itself. Originally, it was thought that the Cambrian explosion was the event where all phyla—the internal body plans of animals—evolved. Now, thanks to the molecular clock discussed in the previous chapter, we know that the phyla evolved earlier in traditional branching manner, and that only their hard parts evolved in a single event—the Cambrian explosion.

Of all the ideas that fell foul of timing problems, the most hyped was the Snowball Earth theory. There were times when the earth's temperature fell so that the polar ice caps extended all the way to, or nearly to, the equator. The earth became a giant snowball. Around 2001 it was proposed that the melting of the snowball caused the Cambrian explosion. This was argued so strongly, in fact, that when I arrived at the Burgess Shale quarry myself that year, I was advised not to concern myself with finding alternative hypotheses because this was *the* answer. We have since established that the last Snowball Earth event ended 115 million years before the Cambrian explosion, and the million-year-long Glaskiers glaciation, with a much smaller global impact, ended some 60 million years before the grand evolutionary event. As I briefly suggested in the previous chapter, the melting of the snowball may have played a role in revealing pristine seas in which the Ediacarans could evolve. But for the Cambrian trigger, a new explanation was needed.

In 2007 Harvard University paleontologist Charles Marshall set to work on the plate of spaghetti that was the list of theories offered to explain the Cambrian explosion. His was a process of elimination. He came to the conclusion that the answer lay not in what happened in the physical environment—volcanic activity or mineral changes—but something that happened in the *biological* environment. The trigger must have been a change in the animals themselves, Marshall concluded. The only theories that held up were the predator hypothesis and the light switch theory.

The predator hypothesis, posed at different times and in different guises, is all about the changes that would occur if a predator were thrown among defenseless animals. Of course, one would expect something dramatic to happen—plenty of feasting, followed closely by new evolutionary innovations. This logic is simple yet sound. But let's think about the predatory animal itself. It should be speedy, agile, and capable of killing and consuming large quantities of food. It also needs to detect and track its food, and to make sense of all the information it absorbs. It needs to make decisions on what to do. It needs a big brain. Brains probably evolved alongside the eye—to exploit the new information that the visual organ was ever more providing. When the eye first evolved and saw the potential around it, its host animal's body and brain would have quickly evolved to catch up and realize the potential. In other words, the new model predators were likely *consequences* of the first eye.

Does this analysis ring any bells? It's as if we were discussing those first trilobites covered at the beginning of this chapter. Now we have arrived at the light switch theory.

The Light Switch Theory

Natural history museums display an array of colors, exhibited for example by birds and insects. Also demonstrated are the variety and incidence of eyes, and physical adaptations to a diversity of lifestyles. When the information on offer here is combined with that from the fossil collections, the workings of today's ecosystems can be applied to ancient environments.

We have learned that eyes, followed closely by predatory bodies, evolved for the very first time around 521 million years ago. These facts are enshrined in the fossil record; the important point is the significance given to them. Now the light switch theory can begin to unfold. Simply, the theory holds that it was the initial introduction of vision in animals that caused the Cambrian explosion. Vision was introduced with the evolution of the very first eye, capable of producing visual images. The view of animals that lived just prior to this, as produced by the most sophisticated light receptors of the time, amounted to no more than a blurred light field. However, the same view around 521 million years ago, again as an inhabitant with the most sophisticated light receptors of the time would have perceived it, would have revealed all the animals surrounding it in glorious detail. As a lens—or rather, the tuning of focus—evolved within the eye, the organ would have made sudden and dramatic leaps in efficiency in a way that other sense organs could not achieve.

The difference between the two perceived images described above is comparable to the difference between what we see with eyes closed and open respectively. Standing in a field, with our eyes closed, we can quickly find the direction of the sun, but we cannot pick out, specifically, food in the form of other animals or plants. So, using sunlight only, animals before 521 million years ago could have known which way was up in the water column, but they could not have found a friend or fled a foe. However, in their favor, a potential predator could not have found them either. Hence there were no strong selection pressures for such early animals to become adapted to light, which was to become the most powerful stimulus of all—with the evolution of the first eye. In fact, light became the most powerful stimulus of all almost overnight on the geological timescale.

With our eyes open, suddenly we see the world completely differently. We can see food from some distance, although we can only smell it if it produces a smell, hear it if it produces a sound, and touch it if we are very close. So, 522 million years ago, not releasing certain chemicals or producing sounds was enough to avoid a potential predator, unless you bumped into it. But a million years later, life was effectively lit up. The light switch was turned on for animal communities, and for the one and only time—it has been left on ever since. With our eyes open we see the size, shape, and color of animals, and we also see how they behave—we can judge how fast they can move and whether we can catch them. All of these qualities of animals suddenly began to matter around 521 million years ago.

At that very moment, all animals had to become adapted to light, or rather vision. Sounds and smells can start and stop, but an animal will always be leaving a potential image in its environment because of the constant presence of the sun's rays; and so the race to produce adapted images began. Some of the wormlike forms began to display armored parts or chemical defenses with complementary warning colors. Others evolved camouflage shapes and colors, or improved swimming capabilities to outmaneuver a pursuing enemy. Another choice was to opt out of the visual environment and evolve bodies capable of burying into rock crevices or the seafloor—out of sight, out of mind. All of these adaptations created a new type of evolutionary arms race. Anything that did not adapt would have fallen afoul of natural selection—it would have been eaten or starved, and either way taken out of the gene pool. But after the initial chaos, further adaptations would become gradual—evolution would settle down to its habitual pace. The Cambrian explosion was over.

So the lights had been switched on for animal life. Remember that sunlight did exist before this event—recall the account in earlier chapters of the origin of the universe and the sun, and God's first command in Genesis, "Let there be light"—but it could not be used by animals to find or recognize each other. Now, if the first eye is added to the geological timescale, the order of events becomes the introduction of vision, first, followed closely by the Cambrian explosion, second. But followed how closely, precisely? If the light switch theory is to hold up, it should be *very* close. Only fossils could answer this question.

It was the age, condition, and diversity of the Burgess Shale fossils that led to the formulation of the light switch theory. These famous fossils represented a whole community of Cambrian animals, providing a rock photograph of community life at that time, with all manner of lifestyles in the shot. That time, though, was 508 million years ago. Yes, the Burgess Shale fossils do inform us that hard external parts had evolved in at least many animal phyla during the Cambrian—but some 13 million years *after* vision had been introduced. The evidence fit, but there was something of a gap between the first eye and the Cambrian explosion event. To strengthen the light switch theory, that gap had to be filled. The search for more fossils continued.

Some fossil "shells" were found that even predated eyes. But these turned out to have been either soft-bodied animals that hardened when they died—altered by unusual preservation conditions—or experiments in exercising biomineralization on a small scale, of the kind you would expect to see as evolution progressed through random mutation (remember, sponges had long been exercising biomineralization). Either way, another, older Burgess Shale—a community of animals fully dressed in skeletal parts—was still needed.

Actually, the Burgess Shale was originally dated at 515 million years ago. But recently the dates of the Cambrian period have been given a makeover. Special attention has been given to radiometric dating of the Cambrian rocks and we now have a very precise sequence of animal community types in geological time, with dates accurate to within two million years. The Burgess Shale is now placed at 508 million years ago, which is the date I have used here.

In 2004 paleontologists from Britain, China, and Sweden published their combined reports of expeditions to the hills and lakes of Yunnan province, in southwest China. Their book revealed a treasure trove of phyla complete with their skeletons, just like those of the Burgess Shale. These were the "Chengjiang" fossils, originally discovered twenty years earlier by Chinese paleontologist Hou Xianguang in the Chengjiang region of Yunnan province. This area had always provided rich pickings for paleontologists, but never before had fossils of such significance been uncovered. In the early twentieth century French geologists had found a plentiful supply of trilobite fossils in eastern Yunnan, and the extent of the Cambrian area was further mapped out in the 1930s and 1940s. But in June 1984 Hou Xian-guang, then a member of the Nanjing Institute of Geology and Paleontology of the Chinese Academy of Sciences, set to work in the region looking for shrimplike arthropods. Eventually, mudstone blocks were dug from Maotianshan (Maotian Hill), six kilometers (a little under four miles) from the county town, by a hired farmworker. On July 1, 1984, one of the blocks was split open to reveal a shocking, semicircular white film. This was instantly recognized as the "shell" of an unknown crustacean—the hard, shieldlike part that covers the head.

Then another specimen was found, this time with limbs attached, followed by another, and another . . . and gradually the reality sank in. Hou Xian-guang had discovered another snapshot of life in the Cambrian—another "Burgess Shale." Indeed, the Chengjiang fossils have emerged to rival those of the Canadian site. Since those first specimens, we have known that they are preserved in glorious "Burgess" detail so as to reveal their finest, needlelike spines. But the compilation of the more recent Chengjiang finds in 2004 disclosed that they also match the Burgess Shale fauna for diversity. And there was more exciting news to come. The new, accurate date given to the Chengjiang fossils was 519 million years old. These fossils represent an end point of the Cambrian explosion—which we now know took place *just* after the introduction of vision. This was the evidence we needed.

So what animals are represented among the Chengjiang fossils? They may belong to the same groups of animals that exist today, but at first sight they certainly don't look the same. Actually, like those of the Burgess Shale they appear as the most bizarre life forms one could ever imagine. *Choia* is a sponge, but seems like a tiny traditional Chinese hat with needles radiating from the edges. *Xianguangia* is a relative of today's jellyfish and corals, but looks like an octopus stuck headfirst in a beaker, with tentacles waving above. *Microdictyon*, a "velvet worm," appears as a caterpillar walking on long legs with rows of shields protecting its body from attacks from above. Then there is a whole assortment of worms from other animal phyla with spines projecting from various parts and into different directions. Again, even the tiniest spines, thinner than a human hair, are preserved in exquisite detail although, in the manner of the Ediacarans, the soft parts were represented in the rock too.

The arthropods, to which crabs and flies belong, are well represented in the Chengjiang fossils. Some are shaped like speed-cycling helmets. But one of the most bizarre is *Occacaris*, which can only be described as a swimming pea with a pair of beady eyes, bull's horns (antennae), and a pair of anthropomorphic "hands" and a fish's tail protruding from its two-part shell. Even so, on closer inspection correspondences could be found between all of these and today's fauna. It was the hard parts that gave them away.

Enigmatic forms also exist, though, so peculiar we cannot even guess to which phyla they belong. One fossil is simply a ball of spines; another is a vase-shaped pincushion. Then there is a discus with a feather attached, a half-worm-half-squid, and also a badminton shuttlecock with a chameleonlike "tongue." *Anomalocaris* appears like an armored cuttlefish, with just two spiky arms, but is unusual also for its size—this animal grew to well over a meter, as opposed to the meager centimeters of most of its neighbors. With its large eyes and shape suggestive of speed, it would have been top of any Cambrian food chain. More recently, though, this animal, also found in the Burgess Shale, has been linked with the velvet worms and the arthropods. We shall see.

The rather fishlike *Myllokunmingia*, on the other hand, *can* be placed into an animal group, and one close to home—the chordates. This is in fact the first known of the chordates, *the* ancestor of humans. Its relative of some ten million years later, *Pikaia* from the Burgess Shale, occupied the finale of Stephen Jay Gould's groundbreaking 1989 book *Wonderful Life*. If this animal had failed to make it through the Cambrian and start a new branch of the evolutionary tree, the story goes, we would not be here to examine its fossils today.

As I have said, the Chengjiang fossils, laid down at the very end of the Cambrian explosion, tell us precisely when this event took place. Now, we know that the only animal with hard parts to predate evolution's Big Bang is the trilobite—that is, the trilobite with eyes: life's first eyes. And now, armed with improved, accurate dates for this trilobite, we know it predates the finale of evolution's Big Bang by just a million years or so. That's how long the Cambrian explosion lasted: the time it took for the worms and globular bodies to evolve their hard parts.

Since the Cambrian, vision has continued to tower over the laws of life. Today, although eyes occur in only six of the thirty-seven phyla, over 95 percent of all multicelled animal species possess them. That's how successful eyes have been, making the Genesis author's choice to highlight the eye in his minimal narrative all the more fascinating. Having said that, his choice to fill the step that followed the eye with a diversity of animals in the seas is equally uncanny, considering what we know now about the Cambrian explosion.

Walk into a field full of animals today and very few can be seen—life is adapted to sight and many animals have developed a natural camouflage to protect themselves. Protect themselves from *sight*, that is; not just from the simple light receptors that existed before the Cambrian, which could not form an image. Vision is behind the food webs of today—there would be no dynamic predation without it. As it was, the earth became filled with teeming myriads of different species as the lifestyle options suddenly exploded.

Some researchers have suggested adding other senses to the mix, so that it becomes the *sensory* switch theory. Today, as I have said, we live in a world full of adaptations to vision, adaptations found even within the small proportion of species that do not possess image-forming eyes. But living animals are also well adapted to taste, smell, hearing, and touch. Nonetheless, vision is the only universal sense in that the stimulus—light—exists in almost all environments, and so, wherever you are, you are leaving an image on a retina somewhere, like it or not. Yes, some animals are camouflaged and can't be seen; but even they are leaving their image on retinas, it's just that the image doesn't stand out—the onlooking animal does not know it is seeing it. This is not the case for the other senses. If you don't move, you can't be heard, and if you don't emit the right chemicals, you can't be smelled.

The big difference, however, is in how the sense organs made their appearance on earth through time. Most senses staggered into the history of life; it is only vision that entered at one, precise moment—in this case with the evolution of a lens. And that is the requirement to trigger an explosive event. That moment—the moment the eye evolved—just happens to correspond with the beginning of evolution's Big Bang.

A critic's view was provided by Timothy Gawne of the University of Alabama. Dr. Gawne challenged some ideas given for why the eye (and in particular a *pair* of eyes) should evolve, but at the same time admits, "The notion that once one animal had functional vision, then every other critter had to adapt, is quite powerful." The new evidence from China shows that they did adapt, immediately. Whichever way you look at it, eyes were monumental in shaping life on earth. "As I see it . . . a predator (a trilobite) evolved for the first time a working visual system, and was so successful that the other organisms had to evolve hard parts to survive, which produced the so-called Cambrian explosion," wrote the late Francis Crick, the codiscoverer of the structure of DNA. "[The light switch theory] arguments seem very plausible to me."

The previous chapter highlighted the statement "Of the five senses, the most important is sight," an assertion to which this chapter has given support. The totally new weapon introduced on earth 521 million years ago just happened to be the most powerful of all. And it was introduced with a bang. Just as T. E. Lawrence's new strategy defeated the modern army he faced, so the eye completely wiped out a whole fauna and paved the way for new innovations that, like the eye, would remain ever after, battling it out in a never-ending arms race.

Further Burgess Shale–type Cambrian fossils have since been unearthed in other parts of the world, such as Sirius Passat in Greenland, dated at around 518 million years ago. Very similar species have been found in each locality around the world, although never in such variety as found in Canada or China. Still, this is a wonderful independent test of the dating system. That similar species are found in unrelated rocks of the same age tells us that we have got the science right. Since that age is over 500 million years ago, it also tells us that the Bible's version of the creation can't be literally correct. The earth is not a mere six thousand years old.

The scene for all evolutionary events covered so far in this book is the sea. Life at the stage we have reached as revealed by science—around 508 million years ago—was exclusively marine. It was also heavily influenced by sunlight. Again, Genesis is correct. Again, one would not expect the writer to have got it right. He writes about the life in the sea, when he is landlocked in his biblical town, and probably knew little of the diversity of life on earth other than those slim zoological pickings surrounding him. The importance of the Genesis author's identity is becoming evident, and we shall examine this question in detail in the appendix.

The term "bring forth" used in this stage of the creation account of Genesis 1 is also interesting. There is a possible analogy with natural selection here. Why would a creator pronounce: "Bring forth the animals"? Surely, if He were creating them himself, he would simply present them: "Here are the animals," He would declare. To cry "bring forth" is to summon a process that will lead to the diversity of animals. Evolution is such a process. The ancient Israelites knew of nothing remotely similar to evolution—not even an alternative process that could have explained why a writer would put these words into the mouth of a creator. Why the author of Genesis should use the term "bring forth" is nothing but a mystery—unless, of course, he had received word that this is the way it really happened. Again, the creation account of Genesis 1 is looking oddly accurate.

Still, as mentioned already, the really great parallel between science and the Bible in this case is the implication that an explosion in life followed the evolution of the eye. There were no clues to be found anywhere in ancient Israel to suggest such a thing, and that the author of Genesis wrote it in this way is rather scary. The eye was a completely novel mechanism thrust upon animal life. The soft-bodied forms of the early Cambrian period lacked any form of defense against the newfangled weapons of vision-enabled predators. For the early Cambrian animals retreat was not an option—they evolved rapidly and radically.

Now we have reached the late Cambrian era. Animal life has flowered and the framework for variety is firmly in place, with all the phyla having evolved, acquired their hard skeletons, and developed adventurous lifestyles. Next, each phylum would explore all the possibilities available to it for new skeletons and lifestyles. One phylum in particular would succeed disproportionately. The small fragile chordates of the Cambrian, those that somehow avoided the formidable arthropod predators, would change a lot. In the following chapter, we will find out how.

Just as depicted on the first page of Genesis, so far we have traveled through time from light, in the formation of the universe including the sun, to seas, in their appearance on earth, to plants, in the earliest life forms, to light again, in the evolution of vision, and now to the variety of creatures that evolved in response to the eye. At this point in the story of the earth, life is developing at a faster and faster rate. Remarkably, so far the Genesis narrative has been confirmed by modern science. Will science and the Bible continue to agree as the story of life continues?

Life Unfolds

"God created great whales, and every living creature that moveth, which the waters brought forth abundantly after their kind."*

The next stage in the sequence of events described in Genesis involves masses of marine life emerging from the seas. Again this is represented using the concept of "bringing forth," suggestive of evolution, as explained in the previous chapter. Does the science agree with the biblical ordering of things? So far, following life's prehistoric journey, the animals from all phyla had become equipped with the characteristic hard parts or skeletons that have been retained until today. At 508 million years ago, the shrimps, snails, and urchins had their shells, and the worms had their spines and shields. But still, collectively, they were a shadow of life's cornucopia today. There were few members of each phylum, and those few were small.

This, the sixth stage in the creation account of Genesis 1, speaks

*The most recent translation (*Today's New International Version*, 2004) gives: "God created the great creatures of the sea and every living and moving thing with which the water teems."

of the "great whales" or the "great creatures." It also employs the adjective "abundantly" (alternatively rendered as "with which the water teems"). Clearly, this stage is describing the further flourishing of life that resulted in the range of living things we are familiar with today—their variety, sheer numbers, and size. The "great" in Genesis means *big*—that, along with the finer branching of the tree of life, is the key concept of this sixth stage. The previous stage concerned the preparation of the animal bodies; this chapter will consider how they fared through geological time, from the end of the Cambrian period until today, and how they grew to enormous proportions. It will trace *how* some of the animal phyla scrambled for their places on the evolutionary tree, now that soft bodies were out and hard, skeletal parts were here to stay. It will also consider how this scientific knowledge penetrated the church's defenses.

In 1957 *Sputnik* became the first man-made structure to be launched into outer space. This Soviet space capsule had a significant bearing on the fate of evolution in science education in the United States.

Since the beginning of the twentieth century, numerous attempts at censoring biology textbooks had been made in the U.S. courts. Some cases were successful, particularly in the South and in parts of the West. As evolutionists continued to present their reliable, scientifically validated data, religious factions tried all sorts of tricks to undermine evolution. Some were dishonest and all were devious. In particular, religious ideas—principally the actions of a Creator God—were cloaked in various pseudoscientific guises such as "intelligent design," which will be covered in the final chapter. In essence, religious beliefs were presented as scientific theories to rival that of Darwin and Wallace, and it was argued that as such they should be taught in school science classes, regardless of their clear bias in favor of divine creationism. And this is where *Sputnik* comes in; for it represented a Soviet lead in the space race, which fueled support for genuine science in the United States—that which might push forward the boundaries of knowledge and technology. The U.S. National Science Foundation sponsored textbooks written by professional scientists. Despite protests from Christians in the United States and overseas, American high schools began teaching evolution, even in the South.

Eventually, the space race would even combine with evolutionary research in the search for life on other planets. Scientific work that pinpointed the precise conditions necessary for the origin of life on earth gave astronomers something to look for in outer space—a planet conducive to life. As signs were found of extraterrestrial water—that essential ingredient in life's recipe—many scientists began to question "where" rather than "if" Martian existence would be located. They were not talking of little green men, but rather of bacteriumlike cells. This was a serious business. Some factions argued the case for convergence—that there is only a limited number of ways in which the problems that face life can be solved, and consequently that these solutions reappear independently in life on different branches of the evolutionary tree. From there, it was suggested that life would always take the same route, and inevitably result in humans, although one has to say that religion lurks here in the idea that if God makes life, He makes humankind. In any event, schools were asked to make their own capsules to be launched into outer space, filled with objects that could convey, in universal language, the manner of life on our fair planet.

The Ark of Evolution

First on every list was a map of where the earth sits in the universe. That apart, most capsules were crammed with items of modern technology, and some with objects of recent social history. But I propose a different approach. How about loading a capsule to best represent the earth with the story of how animal life came to be, telling the tale through key relics from our fossil record? Animal bodies entombed at various depths in our fossil column could provide a self-explanatory history of animal life on earth, all packed into an "Ark of Evolution." In this chapter, as a means of explaining what did indeed take place on earth after the Cambrian period, I will suggest what those key relics should be. This account of the transition of life can then be set against the sixth stage in the creation account of Genesis 1, to determine whether the biblical description is appropriate in what it says and where in the sequence it comes.

The Ark of Evolution should have a foldout timeline to which fossils can be attached at appropriate positions. We must begin our sequence at the start, with the early life represented by fossil cyanobacteria, soft-bodied Ediacarans, the first trilobites (with eyes), and the hard-shelled but diminutive Chengjiang and Burgess Shale fossils. Actually, "early" is not a suitable adjective, since the timeline begins with the formation of the earth at 4,600 million years ago and my chosen Burgess Shale fossil—perhaps the enigmatic *Hallucigenia*—is already positioned as far along as 508 million years ago. But, as I mentioned in the previous chapter, a lot has happened in the most recent 500 million years, even if that amounts to only the final one-eighth of life's history on earth, and all within the thirty-seven phyla already in place. So what fossil should I choose next?

In the 1830s two British geologists discovered groups of fossil animals, roughly 400 to 500 million years old, in the south and north of Wales, respectively. Adam Sedgwick dubbed the period of the fossils he had uncovered in north Wales the "Cambrian," after the local Cambrian Hills, while Roderick Murchison named his period the "Silurian," after a tribe of ancient Britons. The two periods overlapped, although the two names stuck in the geological timescale. At the time, each of the two men wanted to claim his fossils as unique and as the oldest evidence of life. Murchison proved more persuasive through his 1839 volume *The Silurian System*, a finely detailed and authoritative thesis. Yet the argument was only definitively settled in 1879, when a third geologist, Charles Lapworth, devised an additional period to bracket the two together, the "Ordovician" (after the Ordovices, an ancient British tribe in northern Wales). Although all three geologists were without the accurate dating techniques of the twentieth century, and the dates of the three periods have changed since, the labels remain attached to three consecutive periods in the earth's history—Cambrian, Ordovician, Silurian. The Cambrian period we have already discussed; the two following periods are where we should look for the first of the post-Cambrian fossils.

Searching for a symbol of the Silurian, dating back to some 425 million years ago, I was first tempted by the arthropods called "eurypterids," scorpionlike hunters, up to three meters (almost ten feet) long, of the ancient seafloors. Next I pondered the first fishes that evolved from the *Pikaia*-like animals of the Cambrian—although then I decided to save the fishes for the next stage in life's history. No, here I will choose the orthocone cephalopods, ancestors of today's squid, which exploded in numbers during, and are so emblematic of, the Silurian.

There's promise for the future public relations of the currently overlooked Silurian period. That promise is down to the work of the Siveter brothers, Derek and David. That the Siveters derive from the Dudley area of England, west of Birmingham—a stronghold for Silurian fossils—is perhaps a coincidence, but they certainly have breathed new life into the Victorian cases that currently shield the fossils of the Silurian and Ordovician languishing in some of the dustier museums. In the nearby English county of Herefordshire, the Siveters and their colleagues recently discovered fossils that at first sight appeared as squiggles and random shapes rather than animals. That was because the diminutive bodies had been preserved in three dimensions, and in any one section—the way a fossil is revealed at the surface of a rock—the appearance is meaningless. Consider embedding a geological hammer in clay and, when the clay has set, making a random slice in any direction through the whole block. The section revealed would be unlikely to look anything like a hammer. Well, when the Siveters realized that they faced such a problem, the solution became obvious to them.

The Siveters and their colleagues sliced Silurian rock nodules from Herefordshire into ultrathin sections, and pieced together photographic images of the fossil slices using a computer. The computer reconstructed a three-dimensional image of the entombed animal that could be rotated in any direction. Fabulous creatures from an underappreciated era emerged as if they were alive. They included animals with shells such as relatives of snails, shrimps, sea urchins, and sea spiders, although with rather weird and wonderful profiles. Quite rightly they attracted the attention of paleontologists and biologists, even making the pages of the top scientific journals—a first for a Silurian animal. But then these were special.

Nevertheless, despite the modern kudos given to the Siveters' animals, for my choice of the Silurian icon I am drawn to those woodcut figures of Murchison's original nineteenth-century tome, and one animal that captured my imagination as a boy—the "orthocone."

Murchison drew the elongated cones that were the orthocones' shells. Since then, the living part of the animals has been drawn in. At the entrance of the cone projected the head, siphon, and arms. If this sounds vaguely squidlike, that's because the orthocones belonged to the mollusk phylum, which includes snails but also the cephalopod class, incorporating the octopus, squid, and nautilus. The nautilus may be less well known, but is most similar in appearance to the orthocones and their relatives in the Silurian—the ammonites. The nautilus has a spiral shell, like the typical ammonite, and its arms are free to wave in the water, and to grasp prey. Its siphon also opens beyond its shell to jet-propel it through the water. Another notable hard part in cephalopods is the beak—a beak like that of a parrot—to tear apart its prey. In fact, today's nautilus provides a glimpse of the body that projected from the orthocone's shell, since the two are closely related.

Still, despite those similarities, the orthocone would have appeared totally alien to us—a giant tubelike cone with pinhole cameras for eyes and several space-invader arms gesturing in the water. And some orthocones were big—up to ten meters (about thirty-three feet) long, the size of a pine tree trunk! They even gorged on those monster scorpionlike eurypterids. But the giant orthocones are way too big for my Ark of Evolution, so I will select a more manageable shell, a pencil's length: that of an *Orthoceras* that is commonly found for sale as jewelry or in those polished, black rock coffee table tops, where the white fossil shell appears particularly handsome. These fossils also reveal the internal chambers that divide up the length of the shell—once filled with gas to adjust the buoyancy—and the central channel into which the animal's body extended.

In this chapter it is important only to gain a sense of how animal life evolved following those diminutive forms of the Cambrian. Examples of animals are given to provide a flavor of the changes that took place through evolutionary time. But remember that the major branches on the evolutionary tree had all grown by this time, and we are following the fortunes of just some of those branches through the remaining geological time. Our purpose is to make comparisons with the stage in the creation account of Genesis that implies giants and finer variations within each branch of the tree, along with the general thriving of life across the globe. We must judge whether that appropriately sums up what really happened during life's story.

Certainly, the orthocones grew to be giants. But the reign of the orthocone and other cephalopods came to an end in the period to follow—the Devonian, named after the county of Devon in England. The fishes that began to diversify in the Ordovician and Silurian suddenly took the evolutionary reins and the orthocones were left in their well-adapted wake. Orthocones were unwieldy swimmers, although that hadn't mattered in the Silurian since they fed on arthropods that crawled along the seafloor. But suddenly it did matter. In the Devonian the bony fish and early sharks developed into monsters of speed—and again, among them were some *enormous* monsters.

The dreadful Devonian, some 380 million years ago, provided a remarkable contrast to earlier periods. This was the age of fishes, which by now were equipped with jaws. The food web well and truly moved off the seafloor and expanded upward: Evolutionary selection pressures tended to favor increasingly bigger predators, or more efficient eating machines. There had been big organisms before, like the giant orthocone, but in the Devonian period life's giants made the seas their own, and from now on there would always be colossal creatures of one type or another. Not surprisingly, the orthocones themselves were reduced to a few smaller varieties that could keep a low profile. But the complexity of the food web demanded more sophisticated developments as well as simple increases in size. In order to find a place within that web, bodies would require speed and agility, acute senses, and more complex, even clever, behavior. Still, the author of Genesis appears spot on, listing giants as appearing in the stage of creation that followed the acquisition of skeletons.

My own experience with Devonian fishes was as a researcher at the Australian Museum in Sydney. In 1956, near the Australian town of Canowindra, 300 kilometers (about 185 miles) west of Sydney, road works had begun. As machines smashed the rocky ground, one slab that broke away—the size of a door—was seen to be covered in strange impressions. Fortunately, though not giving it a great deal of attention, the workmen were curious enough to put it to one side. A local beekeeper appeared at the scene and recognized the marks as fossil fishes. He contacted a paleontologist at the Australian Museum, and the slab made its way there—where I first saw it. It remains on public display at the museum today. The fossils are of Devonian fishes, all of precisely the same age. These fish once lived together in an ancient pool of water that gradually dried up. As it dried, the pool became smaller and smaller, and brought the fish inhabitants to the same spot, where they drew their last breath. Final sediment influxes buried the carcasses and they became fossilized as a mass grave, fixed forever in the new rock.

In 1993 the new fish paleontologist at the Australian Museum, Alex Ritchie, revisited that grave, equipped with a twenty-two-ton excavator provided by the local council. And his long search paid dividends. Some seventy tons of rock slabs were excavated, all covered in the lucid fossils of Devonian fishes averaging an arm's length.

The fossils are dominated by "placoderms"—armored fishes. These are the reason why I termed the Devonian "dreadful." Their appearance was simply terrifying. The most alarming of all was *Dunkleosteus*, which grew to around ten meters (over thirty feet) in length; with a great girth to match, it was the size of a small bus. Like all placoderms, *Dunkleosteus* was covered in bony armor. Separate bony plates jigsawed together and surrounded the fish's body to make it impregnable to other *Dunkleosteuses*. The really frightening part was the head, and in particular a vast jaw with bony, daggerlike teeth that operated like scissors to crack open the armor of smaller relatives as if parting a nut from its shell. The skull is often displayed in natural history museums and portrays a sense of the drama that must have unfolded in this period of life—somewhat in the manner of Tutankhamen's funerary mask. One views its armored eye nervously. *Dunkleosteus*, nonetheless, did not appear in the Australian site, though it is found in fossil form today in Europe, northern Asia, and North America.

Clearly *Dunkleosteus*, or any defining part of it, will not physically fit into my Ark of Evolution, so for our next showpiece I will opt for one of its smaller relatives at Canowindra—*Groenlandaspis*, which though less massive still carries the unique characteristics of fishes of this age. *Groenlandaspis* is a placoderm with a fully armored head, bony arm projections, and a sharklike top fin attached. It is this part of the body that will represent the Devonian in my Ark of Evolution. It is a little bigger than a hand, but rather elegant, thanks to its streamlined form, and with a touch of mystery about it, since it resembles nothing on earth today.

The two hundred fish-covered slabs retrieved from Canowindra in 1993 are today held at a new museum opened on their burial site, with a viewing platform from which visitors will be able to observe future excavations. The site has been listed as part of Australia's National Heritage because of its international scientific importance. There are almost certainly many thousands more specimens to uncover, probably including hitherto unknown species. Interestingly, one other fossil fish that has already been discovered at Canowindra is a lungfish. Just three groups living in South America, Africa, and Australia represent lungfish today. They are lobe-finned fish, which, as their name implies, have a muscular lobe at the base of their fins. It was a lobe-finned fish that at some point in the Devonian crawled out of the water and on to land. For some time this was believed to have been a coelacanth, a fish that captured the public's attention in the twentieth century thanks to the discovery of a living specimen.

The coelacanth was well known from fossils, but the youngest was some 120 million years old. The group was believed long extinct until, in 1938, a recently dead specimen was found on sale at a local food market in South Africa. It had been fished from deep waters, and today we know of other sites between Madagascar and southern Africa where this fish can be found living. Recently, another was found for sale at a food market, this time in Indonesia, probably representing a second living species.

The coelacanth was hailed a living fossil because of its close resemblance to similar species known from the fossil record. Actually, no species is so ancient, and the coelacanth had changed over the eons, albeit changed very little, making it a good representative of ancient times. But what really excited the public were its bony fins. They were believed to be prototypes of our arms and legs. The story went that the coelacanth hauled itself onto land with these fins, which evolved further to become walking limbs. This story captured my attention as a boy, and I still wonder at coelacanth specimens preserved in museums today. But—alas for the myth, and the coelacanth—the lungfish would have been a better choice as the hero of this tale. For the lungfish has not only bony fins but also, as its name suggests, a lung to breathe air. This lung probably evolved to enable the fish to breathe air while occupying stagnant waters lacking in oxygen, an environment that represented a vacant niche. Evolution does tend to exploit any unoccupied potential home.

The animal that evolved from the lobe-finned fish was possibly *Acanthostega*, a giant salamanderlike animal—an amphibian, belonging to the group that includes frogs and newts today. Changes had occurred to its skeleton that favored walking over swimming, although this animal remained aquatic. Indeed, that first animal that crawled on to land for a fully terrestrial life was not a fish but already an amphibian.

We have now left the Devonian.

The next periods in life's history were the Carboniferous and Permian, centered around 300 million years ago. I will choose my next fossil for the Ark of Evolution from here. And, because from this point I will continue to deal with that single phylum, the chordates, which is generally the most widely known, I will speed up my commentary and skim over the details and stories behind them in order to wrap up this chapter in good time. After all, the creation account of Genesis concerns all the phyla, not just one. Of course, fundamental events did take place in some other phyla, particularly among the arthropods, where the insects and spiders evolved to exploit the land, and the mollusks and various worms that similarly conquered the challenges of breathing air and drying out. Still, we do at least have a Cambrian arthropod and worm, and a Silurian mollusk, in the Ark of Evolution already.

The Carboniferous and Permian periods are best known for the trees that grew in profusion in these times, having since become buried by rock and compressed underground to form coal. Walking between those early trees were two groups of amphibians of interest to us. From one group the lizards, snakes, crocodiles, dinosaurs, and birds would later evolve. The other group would give rise to the mammals. My choice of a representative fossil from this environment is a tooth of the primitive land-living amphibian *Eryops*. *Eryops* looked like a giant salamander with a proportionally large head, the size of a car tire—or, in more familiar terms, like a crocodile with soft skin. Its teeth were numerous and sharp, filling the rims of the strong, gaping jaws, and were covered in enamel. The enamel, which remains on the fossil teeth, tells us that the age of the land-dwelling chordates had begun, for it gives teeth greater resilience and is characteristic of animals that live on land (or whose ancestors lived on land). These teeth would certainly have been useful for crunching those prototype flying insects, much bigger than their relatives today. *Eryops* is our best-known amphibian of the Permian period.

The Permian was separated from the following period, the Triassic, by a mass extinction event caused by a series of catastrophes on the earth 248 million years ago, which wiped out most species of chordates that roamed the land and skies. It was in these periods that the mammal-like reptiles and lizard groups split from the amphibian branch of the evolutionary tree. The prototype lizards were on the small side and rare. The other group, on the other hand, were rather prominent.

The first mammal-like reptiles were the "sail-backs," such as the three-meter- (or ten-foot-) long *Dimetrodon* with a semicircular "fin" projecting from its back along the length of its body. It would have looked like a crocodile-size iguana with a giant fan sticking up from its spine in the manner of a sailfish. But by the end of the Triassic (about 200 million years ago) the tables had turned. The climate became hot and dry, as the single, supercontinent landmass began to break apart. The animal group that would become the mammals took on more diminutive forms and were reduced to insignificant roles in the general ecosystem, while another group of reptiles ballooned into dinosaurs. Large, meat-eating animals patrolled the planet, keeping the small prototype mammals . . . well, small. Dinosaurs would dominate the land for more than 165 million years.

The next two periods in life's history, the Jurassic and Cretaceous, were the age of the dinosaurs. During this stretch of time, from 206 million to 65 million years ago, plate tectonics split apart the continents, although land bridges remained between them, allowing the dinosaurs to spread to all corners of the earth. The dinosaurs, exploiting the warmer climate, diversified rapidly, until the landscape was adorned by their full complement of forms. These included the giant "sauropods" such as *Seismosaurus*, other plant eaters such as *Stegosaurus*, and "carnosaurs" such as *Allosaurus*. Some meat eaters called "dromeosaurs," which included *Velociraptor*, were rather small—some as petite as a pigeon. This analogy is in fact quite apt, since they evolved downy feathers, followed by stronger feathers suitable for flight. This innovation may have helped the small creatures to catch flying insects more efficiently, gradually taking them off the ground. And at some point, fully fledged birds would have graced the Jurassic skies.

The oldest bird fossil we have is from around 145 million years ago. This is the one I would like to include in my Ark of Evolution, since it is the famous *Archaeopteryx*, found in a limestone quarry in Solnhofen, Germany. This fossil sticks in my mind not just because it is so valuable as the first bird fossil—and remains so, despite more recent finds in China—but also for the story behind the specimen itself.

Archaeopteryx was discovered in 1861, which was extremely timely. Just two years earlier Darwin had published *On the Origin of Species*, and the debate was raging over whether all living things had been created together or whether they had each adapted over time. Church officials demanded proof that animals and plants today could be correlated, through species in the fossil record, with more primitive, ancient forms. They wanted to see the links in the chain. So the fossil *Archaeopteryx* arrived right on cue—it had feathers like a bird, yet it also had teeth and clawed fingers, unlike any bird today but rather more like a dinosaur. A missing link had been found. Still, it was not until a dinosaur with very similar although featherless features was discovered in 1968, and Chinese dinosaur fossils with only simple feathers were unearthed in 1995, that the transition from dinosaur to bird could be fully visualized. So the magpie-size *Archaeopteryx* enters my Ark of Evolution to represent the first birds and the Jurassic.

To epitomize the dinosaurs and the Cretaceous I will, predictably, choose one of the largest terrestrial carnivores: the fourteen-meter- (or forty-five-foot-) long *Tyrannosaurus rex*, which lived at the end of the dinosaurs' reign—or rather, to fit into the Ark, one of its nine-inch dagger teeth.

Large, air-breathing reptiles would flourish in the sea but also conquer the air in this same period, in the form of plesiosaurs and pterodactyls, for instance. But it would be all over for them at the end of the Cretaceous period, probably because a giant meteorite struck the planet, with devastating repercussions. The meteorite explanation for the dinosaurs' demise is the best theory we have, but it is still just that, a theory in the true sense of the word: That is to say, it could be wrong, as opposed to "evolution by natural selection" which, as I will argue in the final chapter, is more appropriately regarded as "fact." In general, theories never become facts in science—they explain facts. But since this does not translate well into everyday language, where "theory" is held to be synonymous with "could be wrong," evolution is best described as a fact.

Now we are into the most recent 65 million years, collectively known as the Tertiary and Quaternary periods. With the dinosaurs out of the way, the mammals that survived the great extinction were able to leave the cover of the undergrowth and flaunt themselves as the new top predators and giant herbivores of the land. Winding the clock forward, we see tiny shrews become saber-toothed tigers, giant sloths, and mammoths. Fatefully, one branch of the rodents also evolved to become the primates—monkeys and apes.

Switching to a finer timescale, by 160,000 years ago, in Kenya's

Rift Valley, a branch of the ape lineage had evolved—through leaving its treetop home and exploiting the savanna beneath on two legs—into the modern human. As mentioned briefly in chapter 1, these humans traversed the globe. They traveled south into South Africa and north into the Middle East, from where they spread to northern Europe, India, Siberia, Southeast Asia, New Guinea, and Australia by around 22,000 years ago at the latest.

By at least 10,000 years ago, urban settlements existed in the Middle East, developing into civilizations that rose and fell over the following millennia. By 1400 BCE humans were mining and smelting copper in the Negev and the Sinai. And then, in 1200 BCE, a man named Moses allegedly encountered God. God's word was eventually recorded for prosperity—and, strangely, the first page in the manuscript echoes the complete history of life on earth as represented in my Ark of Evolution. What should the last object included be but a skull of *Homo sapiens*? We have returned to chapter 1.

EVOLUTION AND TIME

When I saw the sights of London's "cathedral of nature"—the Natural History Museum—as a boy, I pieced together my own picture of life through the ages. And I did this through the iconic images my mind selected involuntarily from the multitude of options. These now sit in my Ark of Evolution. At this time, I was not interested in religion, and saw no reason why religion should have a serious impact in natural history. Nor had I been brainwashed by science or family to accept evolutionary theory—it just seemed to make sense. I had no intention either to agree with or argue against whatever the Bible had to say.

Not only was I unaware of the heated science-versus-religion debate, I was far removed from the efforts that had been made in some quarters to reconcile the apparently opposing sides. In 1909, while Charles Doolittle Walcott was busy unearthing the Burgess Shale fossils that were to be so damaging to the creationists' case, the Scofield Reference Bible was published. This was a new version of the King James Bible in which a monumental amendment was made to the book of Genesis in the form of a note added on the "gap theory" of creation. This was the suggestion that millions of years could have passed between the moment when God created the heavens and the earth, and the subsequent six "days" in Eden. This allows for the acceptance of an ancient earth. Unfortunately its appearance was followed by numerous crusades and bills in the United States to reimpose the literal translation of the Bible and deem unacceptable any deviation from the precisely literal reading of scripture. In particular, many antievolutionists abhorred the idea that humans are closely related to apes and other animals. There is a "demoralization involved in accepting a brute ancestry," according to the politician and orator William Jennings Bryan. And, in the backlash against the accumulating evidence for evolution and the ancient earth, as mentioned at the beginning of this chapter, the censorship of biology textbooks in the United States followed.

The Scofield Reference Bible nonetheless had something extremely sensible to say. It put aside for a moment the prejudiced views of social conservatism to reveal a creation account on the Bible's first page free from the shackles of a six-day process. What it left was a series—the same series—of timeless events; and it is these that match the scientific account of life's history. Suddenly all becomes factual.

If I were to boil down into a concise list the description I have given of those objects I placed in my Ark of Evolution, it would read like another Genesis creation account, expanding only on its sixth stage: "The seas carried through time the orthocone's shell in great numbers, until the fishes became giants and relinquished the sluggish cephalopod's rule. Some fish became salamanderlike and crawled out of the water, along with other sea creatures, to blanket the land. In a fight for supremacy, living organisms grew bigger as one creature gave way to the next. Giant lizards then giant rodents ruled in succession. While birds and butterflies claimed the air as their own, some rodents became apes, which in turn gave rise to humans." There are no units of time in this passage, and without our biological and geological knowledge we might superimpose a timescale of seven or eight days on to these sweeping events. If this were the first time we had encountered such a story, to add popular appeal in the manner of all good storytellers we might wrap it up neatly into a romanticized and easily visualized week of activity. An interesting thought.

Cyrus Scofield, the editor of the Scofield Reference Bible, held a strong belief in God, but understood how the Bible had come into existence and how it should be treated. If Cyrus Scofield had read *The Genesis Enigma*, I'm sure he would have found further satisfaction in his metaphorical approach, but also strength in his accuracy—accuracy that was denied by others in the church.

I often sit in Richmond Park in Greater London to watch kestrels hunt mice over an open, well-tended field of native grasses. The kestrels rest in the trees that surround the grass plain, and periodically circle in the air above it, hovering then swooping at awesome speed and with extreme accuracy whenever a rodent is sighted. I contemplate the parallels with the savannas of Africa, where lions hunt zebra and wildebeest in similar fashion. An arms race exists on earth, with the hunter and hunted battling it out, continuously evolving new weapons and counter-weapons. Then I make sense in my head of our evolutionary history—why the predatory squids in the oceans today show similarities with their ancestral ammonites, but lack the parts that became outdated when novel tactics developed elsewhere in the water. The ammonites' large shells, which once propelled the animals vertically through the water at great speed, suddenly became their Achilles' heel when another animal evolved that could puncture it, and so that "cephalopod" branch of the tree of life continued without the bulky shell.

Life is not a great chain of being from sponges to humans, but an intricate tree, with numerous branches existing simultaneously and continuously splitting apart. Sponges did not disappear when the next stage of life—the jellyfish and corals—emerged. All these different groups of animals, represented as branches on the tree of life, have always existed, and many have battled for supremacy in an arms race.

In this chapter we have seen different types of animals fulfilling the roles of predator and prey in a similar environment through time. First the trilobite exploited the soft-bodied forms; then the larger Cambrian creature *Anomalocaris* evolved to give chase to the trilobites; then the orthocones hunted crustaceans on the seafloor—and so on, until today we have kestrels hunting mice, and lions hunting zebra. The point is that since the Cambrian a system has existed with various hierarchies, and animals from different branches of the tree have evolved to fill the various layers in the system. The multiplicity of branches on the tree of life means that competition is fierce, and an innovation in one branch could spell disaster for another that has enjoyed a spell of primacy. Once the dinosaurs ruled, then the mammals took over; but the system or food web worked in the same way.

It was the advent of eyes and sophisticated brains that kicked this system into place. Once vision was in place, there was no going back and no changing the system. And the evolution of vision was one of the stages in the creation account of Genesis 1. It seems that the author chose his small number of stages well.

It is interesting to see how the science of paleontology has progressed since the days of Darwin and Wallace. Not only is the fossil column far clearer, but it can be viewed as broken into sections where animal communities change dramatically. Often as a result of natural disasters on the planet, great extinctions took place, followed by the rise of groups that had previously lived in the shadows of others. In essence, the fossil record tells a story of evolution in fits and starts; the idea of gradual evolution is well and truly buried. Evolution has not proceeded at a steady, uniform pace.

This, the sixth stage in the creation account of Genesis 1, again shows remarkable accuracy when compared with the scientific story of life's evolutionary journey. Here, the Genesis writer envisaged great creatures evolving from those tiny Cambrian forms, eventually making their way out of the sea. Eurypterids, orthocones, giant armored fishes, dinosaurs, and mammoths all emerged from marine ancestors that evolved their characteristic hard parts during the Cambrian explosion. Even the land-dwelling birds of the past grew to be giants. In fact, gigantism happened in most phyla. Certainly these extinct animals could be described as "great," if only for their sheer bulk.

This revered earth has played host to countless ages where many animals have existed and then passed away. Nonetheless, we can employ Darwin and Wallace's evolutionary tree to place these events in context, revealing which are the major branching moments of the tree and which are merely fine-tuning of the branches. The evolution of dinosaurs, in this respect, occupied a level of importance beneath that of the evolution of the thirty-seven phyla of animals on earth, or the moment when these phyla evolved their skeletons and their evolutionary activity soared. Genesis 1, it seems, picked out all the events of highest order of importance, and put them in the right order.

We have passed from the first stage in the creation account on the Bible's opening page to the sixth, and found it all remarkably accurate, as if the modern scientific story of the universe and life were being narrated. I don't know the odds against such a parallel—against making a successful guess at the scientific orthodoxy of three thousand years in the future from a knowledge base of nothing—but they must be extraordinarily long. As I first looked through this sequence of Genesis, I did not think that we could possibly reach the end of an impartial history of the universe and life without finding more than a few obstacles. I thought Genesis would fall at the first hurdle or two. I'm amazed that we have made it to the end unscathed.

In some ways it is sad to see those centuries of knowledge, built from the ancient Greeks up through Copernicus, Newton, Buffon, Linnaeus, Cuvier, Darwin and Wallace, and Einstein being used as a weapon to destroy religion. The question I cannot avoid is: Why? On the one hand there's nature and the scholarly view of how it got here and how it works, but on the other there's the wonder and beauty that somehow make up another "subject" interwoven with the science. Yes, we've gradually built our scientific picture from simple beginnings—and this book has offered a glimpse into this process. Maybe we should never reach a stage where science should put the wonder and beauty—possibly synonymous with religion—to death. Still, to adopt the words of the nineteenth-century British poet Alfred, Lord Tennyson—although he intended a skeptical comment on evolution—Nature *is* "red in tooth and claw." We must accept science.

Moreover, since we have reached the moment of humankind in life's history, this is the end of the journey of energy . . . so far. Energy is something we unavoidably picked up at the very beginning of our investigation into the history of the universe, and it seems to have stuck with us ever since, revealing itself at every stage. Energy was first responsible for the planets and all matter, which was altered into a form having life. Energy has been behind everything, and all through life's history those animals that use energy best—that do things more efficiently—tend to be those that survive to leave a legacy. The cosmological and biological tales have been interwoven through our history of scientific endeavor, yet these two stories only surround a central subject that is energy. They never penetrate it. Energy remains a chasm in our knowledge, and in the Genesis Enigma it has singled itself out as something special. More about this is in the final chapter.

After exposure to the full force of science, I have to admit that Genesis is showing signs of enlightenment. Still, we haven't quite finished yet. There's one more stage in the creation account of Genesis 1 remaining—birds.

Birds

"And God created . . . every winged fowl after his kind."*

The author of the Genesis creation story made another interesting zoological decision after he had seemingly set up most animals on earth—to devote an additional phrase to birds. There is another interesting parallel here between the scientific account of life's history and the creation account of Genesis.

Why a special case is made for birds in Genesis, I do not know. In evolutionary theory, birds are of particular interest because they are an exception to the rule of vision, the sense that caused evolution's Big Bang and presides over the laws of life today. Curiously, birds are our prime living pointer toward the light switch theory for the cause of the Cambrian explosion. That will be the theme of the first part of this chapter, after which the relevance of this seventh stage in the creation account of Genesis 1 within the scientific summary of life's history will I hope become clear.

*The most recent translation (*Today's New International Version*, 2004) gives: "Let birds increase on the earth."

EVADING THE PREDATORY EYE

When I find a sea-smoothed pebble on a beach, with white veins meandering through grayness, it reminds me of the cycles of the earth and of those millions of years of geological activity. When I halve a rock and find a fossil, it reminds me of those millions of years of evolution that have shaped the life we see today. But when I see a colorful bird, I am reminded of eyes, and the impact of vision on the earth in its immense contribution to the evolutionary process. For the reason why so many birds are colorful, particularly in sunny zones, is that they have escaped the rules laid down 521 million years ago by that very first eye. Birds can fly, and so evade the predatory eye. Leaving aside the relatively rare phenomenon of birds eating other birds, they are largely free to flaunt themselves visually.

Whenever I think of birds, the first fauna to come to mind is that of New Zealand, which surely hosted the most exceptional examples. If one had the opportunity to see these in the wild, the large, flightless, egg-shaped variety with beaks the shape and color of strawberries—the takahe—would surely appear just plain weird, while others would be difficult to identify: parrot or crow? Then one would need to duck for cover as the muscular moa—twice the height of an ostrich and several times as persuasive—broke the sunbeams through a forest clearing. These are not our feathered friends, as most of us would know them. But the bizarre birds of New Zealand are largely exceptions to the rule. Most of these eye-catching species were ground-dwelling creatures because the usual predators of birds—those that hunt using vision—were absent in New Zealand. There were no cats, foxes, stoats, or the usual bird-killers and nest-raiders found on other landmasses. Well, not until humans arrived.

The first human immigrants—the Maori people, who derived

from Polynesian tribes of the Pacific—caused some damage to the native wildlife with the introduction of the Maori rat. But the next great influx arrived in tall ships, complete with European rats and cats, brutally cutting short the sovereignty of ground-dwelling birds in New Zealand. Numerous extinctions followed in short order, and although some species still survive, just barely, on the main islands, most are confined to protected areas—glorified zoos—on the small islands.

The moral of this story is that if one is hunted by a predator with vision, one must develop counter-weapons. This usually means either camouflage or warning colors, along with behavior appropriate to the selected strategy. The birds of New Zealand had neither, and suffered the consequences. Most birds, on the other hand, have evolved one strategy or the other—and frequently they go all out to be noticeable. They exploit the complete paint box. It seems only appropriate, then, that one of the pioneers of ornithology came from a divided background of science and art.

John James Audubon possessed both the attention to detail necessary to make progress in science and the hand of an artist. His depictions of birds did not end with visual brilliance, but invoked all the senses with the poetic expressiveness of his accompanying stories. The reader is drawn into a bird's environment, experiencing the way it rummages, or, as New Zealanders would say, fossicks on the forest floor or negotiates the canopy.

Audubon did not have an easy start. He was born in 1785 in Saint-Domingue (now in Haiti), the illegitimate son of a French sea captain. At the age of three he was sent to France, partly to conceal his illegitimacy, and in 1794 was legally adopted. Later he began to paint birds, initially copying the woodcut drawings of Buffon's *Histoire naturelle*. In 1803 Audubon eluded conscription into Napoleon's army by emigrating to Pennsylvania. He infiltrated the

American forests and mountains like his childhood adventurer heroes, and emerged a hero himself. After flirting with conventional employment he accepted his destiny and returned to his inescapable passion: to paint and catalogue birds.

It has been said that to Audubon birds represented a parallel universe to human experience. He used every type of paint available to capture the stiffness yet weightlessness of the feather, and above all the remarkable hues. He captured the velvety green bodies and vibrant yellow and orange heads of the Carolina parakeet, and the blue-green flashes of the grouse's tail. To reproduce these vibrant originals in books, and multiple copies, black and white etchings were printed and then hand-painted with watercolors. Audubon's tour de force, *The Birds of America*, advanced ornithology several stages from Buffon's unrealistic, and occasionally anthropomorphic, woodcut prints (although Buffon's fame was not primarily as an artist).

As his health declined Audubon would move on to mammals, with the help of his sons. But by this time he was already acclaimed as America's first great watercolorist, and a pioneer of art and science. Soon he would also be known as an environmentalist. Still, magnificent as Audubon's watercolors are, they sometimes lack a certain something when compared with the living hues of the models themselves, for nature has more ways than the artist of producing color.

We first picture a palette for feathers, as for all of nature's hues, like that of an artist, with pigments of various shades ranging from introverted pastels to bold primary colors and shocking pinks. All these pigments share a common mechanism to generate their colors—a molecule of some kind that alters the white sunlight that strikes it. A complex collection of atoms devours some of the energy in the sun's rays—converting their light energy to heat energy—and rejects others. What we see are the rejected rays, which are scattered back into the environment. Since our visual systems convert different rays to different colors, effectively some colors are removed from the rainbow that strikes our eyes. With nothing removed, we see the full rainbow, whose colors pooled together appear white (think of the prism, which splits white light into its spectrum, in reverse). If all colors except red are removed, we are left to see red. That's the job of a red pigment.

But bird feathers are not just packed with pigments. They are more sophisticated than that. One other way in which they appear colored is, ironically, through incorporating transparent materials. Yes, *transparent* materials. But how can a transparent material produce color when it is by definition see-through? Well, the material does not pack the feathers as a solid block, but rather is sculpted at the microscopic scale with elaborate architectures, leaving lots of air spaces and surfaces themselves invisible to the eye.

These microscopic sculptures split up white sunlight into its rainbow, and reflect back into the environment, and to the eye, only some of the sun's rays. These "reflected" rays appear a certain color. The rays that are not reflected pass straight through the miniature sculpture, never to be seen. This is where the architecture of the sculpture becomes important. The precise design has a signature colored effect. Some have the effect of a hologram like that on a credit card or a compact disc. But notice how the reflections of these man-made items appear particularly dazzling—metallic, iridescent even. The same effect is also seen in the bird's plumage. Think of the peacock's tail or the hummingbird's. What a wonderfully energetic system. Yes, energy—the same energy that first appeared at the moment of the cosmological Big Bang, recycled so many times over the 13,700 million years since.

Now here's the problem faced by the ornithological artist. He is armed with the full spectrum of colors, but all made only from pigments. Pigments manipulate the energy in sunlight in a very precise way. The hummingbird's iridescence cannot be captured by such means.

As Audubon sought to conquer the palette found in birds' feathers with the artist's paint box, the next ornithological legend went a step further. This painter would invent a new means of applying color to paper—a technique that emulated the microscopic sculptures in hummingbird feathers. He achieved iridescence in his paintings.

Born nearly twenty years after Audubon in 1804, John Gould began life as the son of an English gardener and ended up as one of the most famous ornithologists of all time, influencing even Charles Darwin with his skills in taxidermy, taxonomy, and artistic realism. As an innovator and expert in taxidermy, despite his scant education Gould found a position as the first curator and preserver in the museum of the Zoological Society of London in 1827. It was through this position that he became acquainted with Britain's leading naturalists.

Gould dealt with Darwin's bird specimens collected on the *Beagle*'s second voyage. He gave new identifications to Darwin's original categorizations. Decisively, the Galapagos mockingbirds, which Darwin had presumed were mere varieties of the same species, came to be classified by Gould as separate species, with a relative on the South American mainland. It was these mockingbirds that finally gave Darwin the idea of speciation through natural selection. Actually, Darwin had been a little carefree in labeling his mockingbirds without recording their precise island home. Fortunately, Captain Fitzroy and other members of the expedition had been more diligent, and it was through their specimens rather than his own that Darwin was able to establish that each species was unique to a particular island. They had *evolved* new characteristics to help them survive their unique conditions.

Gould published his works on birds and mammals with entrepreneurial deftness. He enhanced his scientific contribution by the addition of his artistic skills and rolled the whole package into a series of beautiful folio volumes—possibly the most attractive to be found in the literary world. I found a copy of one of them in that same elite antiquarian bookshop in London. In their own time, Gould's works became a must-have for leading figures in society, particularly following endorsement by the Prince of Wales and a place in the Great Exhibition of 1851 in London's Crystal Palace. Gould gathered subscribers, guaranteeing his books' success before they even left the printing presses, and amassed a personal fortune. Much of this he owed to the real breakthrough he could claim on the pages themselves: He had, for the first time, captured the iridescent colors of birds.

Well, to his readers it was for the first time, but he may have conveniently skimped on a proper acknowledgment. William Bailey, a hummingbird specialist from Philadelphia, was a generous man and had shared with Gould his secret of achieving iridescence on paper. Gould did admit to learning of Bailey's technique, but claimed not to have used it. Gould's and Bailey's techniques, nevertheless, were at least very similar. Their chief secret was gold leaf.

Hummingbirds were an infatuation within a passion for Gould. He wrote incessantly of their "glittering hues," "sparkle and shine," and "metallic parts," and described the birds themselves as "living gems." Actually, "living" was wishful thinking on Gould's part, since he worked on hummingbirds for twenty years before setting sight on a living specimen.

Among the hummingbirds, Gould's favorite was *Loddigesia mirabilis*. The *mirabilis* was first found in Peru in the 1840s, and stood, aptly, as the jewel in the crown of the most famous hummingbird collection, that of the British promoter of natural history George Loddiges. The charitable Loddiges allowed Gould access to his collection, but Gould would offer as much as £50 (about US$75)—a small fortune at the time—for another *mirabilis*. Despite all efforts, a second specimen was not found until 1880. Still, Gould was able to project a lifelike image from Loddiges' stuffed specimen.

Among the 360 color plates of his monograph series on hummingbirds, produced between 1849 and 1861 (during which time *On the Origin of Species* was published), *mirabilis* is represented by three imaginary specimens, two in flight and one at rest. Their turquoise wings and emerald throat show a matte appearance, typical of the usual watercolor effect. But there is something extraordinary about their electric blue crests, their green and yellow backs and tail, and the rhombic plates that terminate on two extra-long tail quills. These bear a metallic sheen. The color even appears to fade out as the page is tilted so that one views it edge-on. One gains an impression of movement not usually associated with watercolors. Colors and light beams flash and flicker across the retina. Some aspects of this effect provide a sense of realism.

Gould was careful not to divulge every detail of his secret formula for iridescence, and so we will never know how close his method came to Bailey's. We do know that Bailey began his process by laying down a metallic surface, sometimes using metallic ink and sometimes gold or silver leaf—a common ingredient among the top rank of Bible copyists in monasteries before the advent of the printing press. Gould did at least admit to using gold leaf. Then Bailey would apply a very thin film of gelatin. Finally he would add the colored pigments responsible for achieving the desired hue, which he uniquely ground in honey. Gould evidently employed some form of transparent oil- and pigment-rich varnish. Whatever the case—

coinvention or copying—these recipes achieved both the brilliant metallic glare characteristic of hummingbirds and longevity on the page, since the paint did not crack or scale off. Both Bailey's and Gould's paintings were unaffected by changes in temperature or humidity. But how did they work? The answer is interesting because, like the hummingbirds' feathers themselves, they rely on a mechanism other than the commonplace pigments.

The metal background to the color provided a mirrorlike reflection—just like any metal object, it reflects sunlight as a beam, rather than diffusing it into the environment in the manner of a pigment. Artists' paint appears the same to the eye from all directions, rather than the now-you-see-it-now-you-don't of a light beam. The honey or varnish spaced out the pigments so that the layer remained semitransparent, yet the pigments were concentrated enough to impart a hue to the beam that was back-reflected from the metal background.

In essence, Gould and Bailey achieved the same visual effect as the hummingbird feather, albeit by different means. Hummingbirds rely on neither pigments nor metals to achieve their metallic coloration. Replication of the devices found in hummingbird feathers would have to await the introduction of nanotechnology, an innovation only of recent years.

Actually, nanotechnology has now advanced to the point where we are making microscopic structures with precisely the architectures found in bird feathers and other animal parts. A particular Australian weevil has a color-producing structure identical to that of the gemstone opal—minuscule spheres closely packed together. Today, we can make the same device ourselves. A marine worm called a "sea mouse" has hairs filled with minute tubes (invisible to the naked eye), again all packed together and running along the length of the hair. Light is reflected inside this device to produce wonderful, changing metallic colors. Again, we can manufacture the same device today. And, just as Gould would have done, a contemporary artist—Franziska Schenk (not to be confused with the Olympic speed skater)—is using our factory-made analogues to paint these animals in their original color-producing mechanisms.

I couldn't help but introduce the subject of metallic colors because, like vision and the Cambrian explosion, it is another of my research topics. I use an electron microscope to examine insect shells and bird feathers, since this machine can achieve magnifications unheard of in Gould and Darwin's day. In the beginning, I expected a lifetime of unearthing new microscopic architectures within the earth's variety of animals. Then overlaps became the norm. Hummingbirds have color-producing structures identical to those of the sea mouse, and single-celled life forms produce the same "opal" structure as found in the weevil, for instance. Finally I found the reason for this: Cells were using their internal machinery to make these body parts.

Nature's Architectural Economy

Imagine cells as tiny factories, all with the same machinery. If the machines are switched on, they make their distinctive products, or body parts. If switched off, they stop work. This is why the same color-producing devices persistently reappear in birds, and indeed in animals in general. All that machinery was put in place in the single-celled life forms that lived some 2,000 million years ago, and from which all multicelled animals evolved. It is the genes, of course, that have flicked the on/off switches throughout evolutionary history. But the point again is that the complexity of animal bodies is not so astounding that it defies our understanding and scientific explanations.

In fact, the factories that make animal parts *in general*, from the invisible to the substantial, are ancient—predetermined when the first animal cell came together. And, as expected from this model, animal parts in general are not different in all species. Comparable architectures occur and recur as species evolve through geological time, where the machines within cells are turned on, off, and on again. Once more, this makes evolution far simpler to explain: As a process, it is far less complex than once imagined, and no longer improbable. It is conservative with its limited energy—the wing muscles and the mirrors in the eyes of butterflies are made with the same cell machinery, and so share the same microscopic architecture. The architecture *and* materials are the same in the flexible mirrors that make a cat's eye glow in the dark and in the gas-proof walls of a fish's swim bladder. But then, when something is barely possible for the general cell to achieve—a green pigment, for instance—it is simply not found in any animal (to generate green, most animals are forced to evolve the more elaborate procedure of producing both blue and yellow, which combine). The natural process works without the need for an intelligent force, or Creator. After all, making both a blue and a yellow pigment instead of a single green one is hardly intelligent.

The message surfacing at this point is by now a familiar one: When the biblical text is taken literally, it is left in the wake of an advancing science. But when it is read figuratively, as here in the case of birds, it becomes a great unknown in the way it keeps pace with modern science.

Returning to New Zealand, today we would not find its ground-dwelling birds so outstanding, since most of that unique menagerie I described at the beginning of this chapter has gone. Those introduced foxes and cats saw to that. What are left, among those birds that cannot fly, are the likes of the ground parrot and kiwi, which are not only dressed in camouflage colors, but also exhibit stealthy behavior or have given up living in sunlight and become nocturnal. Those New Zealand birds with working wings, on the other hand, have largely remained safe to flaunt themselves.

Like the hummingbirds, birds of paradise, parrots, pheasants, ducks, finches . . . in fact the males, at least, of most birds, the feathered inhabitants of New Zealand canopies are gaudy. It is always an advantage for an animal, and the survival of its species, to be brightly colored in order to attract a mate, and ultimately reproduce. Usually, in most animals, the disadvantages of also attracting a predator override this; hence the cryptic colors that abound in those that have no escape route. Flight, clearly, is the perfect counter to most predators. Most birds that can fly have an answer to the images they leave on the retinas of most predators.

Birds, then, represent a group of animals that abstain from the general rule of vision. If the diversity of *sea creatures* evolved originally in response to vision, then *birds*, evolving late in the history of animals, serve as a message of the power of light and vision on earth today. They tell us of that great authority of life—the sun—through their brilliant hues, which so many of their fellow animals can ill afford. It was fitting for the author of Genesis to say something special about birds in his creation account.

Having made a case for the deliberate mention of birds as a separate entry in the catalogue of creation, we must consider one further point made by the Genesis author. Before intelligent humans appeared, the author added one last phrase to his creation list that represents an activity in life's history. God said, "Let the earth bring forth the living creature after his kind, cattle, and creeping thing, and beast of the earth after his kind." This must refer to the suite of animals in existence before contemporary humans. That is the culmination of all the nonhuman evolutionary activity to here, which necessarily includes repeats. But one new addition is "cattle"—suggesting the wild cow, which would become domesticated after the appearance of "Adam," or the first intelligent human. There is little to say here other than that this is a fitting precursor to Adam. In life's history book, animals appear at this point that would be exploited by modern humans. In a nutshell, when describing how the planet and life around us came to be, the writer of the Genesis narrative got it disturbingly right.

Human Destruction, Human Knowledge

After digesting this wonderful story of how the earth and life came to be, I'm left with something on my conscience—the behavior of just one species.

Although I am most intrigued by New Zealand's avian fauna, the one bird of the world that first springs to my mind is the great auk. My antiquated family encyclopedia of animals, now in its seventh generation, is titled Knight's *Pictorial Museum of Animated Nature*. It contains an entry on the great auk, which provides a detailed account of the bird's environment and behavior. When, as a boy, I saw a stuffed specimen of the great auk—looking like a large penguin or razorbill—in a natural history museum, I was at first impressed to see the woodcut image finally translated into its 3-D form. Then I was dismayed to read the word "extinct" on its label.

Now, think of all those millions of years of evolution that took place to finally leave the great auk on our planet. Think of all those countless transitions that have taken place from single-celled life, through *Pikaia* of the Burgess Shale and the primitive bird *Archaeopteryx*, to reach the stage of the auks. What a staggering and wonderful journey. Then humans came along some 160,000 years ago, and in the last couple of centuries—ironically, as the application of their intelligence reached new heights—they cut off the majestic great auk's existence in its prime.

This could happen to most other animal species with which we share the planet today. Audubon, while studying his bird subjects, watched the logging of American forests with horror. He commented, on the destruction he had observed to take place in just twenty years, that "although I know all to be fact, [I] can scarcely believe in its reality." Alfred Russel Wallace spent a considerable number of pages conveying a similar feeling as he witnessed deforestation and soil erosion. I have watched with remorse tourist divers in Southeast Asia returning to their boats with armful after armful of "souvenir" coral. It would be extremely sad, and also perilous—given the disturbance to life's delicate balance it involves—if we were to continue to allow irreversible carnage to destroy the natural world. And if we keep going as we are, that is what will happen.

Don't think that the Bible will solve this. No amount of prayer will correct our current behavior toward the environment. And don't think that scripture tells us that everything will be all right—any biblical passages that seem to suggest this should not be read in this literal way. The Genesis Enigma has told us that. If we can take just one thing from science—from all those cases described in this book of how we have advanced our understanding of the universe—then it should be the way that problems are solved through careful adherence to the facts, ingenious thought, and rigorous testing of possible actions. If we are to save our planet's wildlife, and ultimately our planet itself, we must turn to science.

I will refrain from citing global extinction figures—the single fact given above, that since a worldwide trade in hummingbird plumage had begun, only two specimens of the magnificent *Loddigesia mirabilis* were found over the next forty years, speaks volumes. Rather, I will leave an optimistic message: that we can reverse our recent ruin of the environment. There's still time, there's still a vast ocean with currently enough oxygen-producing, microscopic life forms to keep us breathing . . . but we must let go of the literal truth of the Bible before it *is* too late. In 1665 Robert Hooke recorded in his revolutionary thesis *Micrographia* that "The truth is, the Source of Nature has been already too long made only a Work of the *Brain* and the *Fancy*: it is now high time that it should return to the plainness and soundness of *observations* on *material* and *obvious* things." Well, Hooke had a point. If we want reliable answers to our questions, we should turn to the scientific method. To put this another way, the only problems that humans can solve for sure are those that lend themselves to science—that is, those where quantitative data, facts and figures, can be found. But Hooke's was an era of scientific infancy, and those who were collecting and using the facts and figures to formulate testable theories must have been excited at how well this method worked. They must have felt that science was the answer to everything. Indeed, as the scientific decades and centuries progressed, throwing up the discoveries and enterprises I have described in these chapters, Hooke would appear to have been proved right. We did indeed solve problem after problem with science.

Where science advanced, religion retreated. Science's gains were religion's losses. That's because of the way the relationship was set up—God was used to explain the gaps in our scientific knowledge, becoming the so-called God of the gaps. The inevitable problem here is that as the gaps are narrowed, God's influence is diminished. Then a time comes—as it indeed has come—where there's not much left to credit to God. This is where Darwin and Wallace's theory of natural selection caused so much damage—it filled not so much a gap as a chasm.

But maybe we have been looking at religion in the wrong way. If we scrap this model and start over again with our religious thought, we may rediscover God. That could be the conclusion to which these eight chapters lead.

Rather than focusing on all those gaps in our scientific understanding, what about turning to consider that *energy* that has been flowing through the universe and life from chapter 2 to here? Since the cosmological Big Bang, energy has taken a remarkable journey, transformed into atoms that in turn formed rocks, oxygen gas, water, and life. Energy was also responsible for the forces that hold the atoms, rocks, and living bodies together, as well as stabilizing the planets and the universe itself. That all this happened, and that the energy in that first life form transformed itself, via evolution, into fishes, shrews, and finally humans, is rather mind-boggling. One question to be explored in the following chapter is whether these sequences of events were predetermined, in that the universe and life would unfold in the same way if the "experiment" were repeated, or whether human existence is pure chance, and could never happen again. This is the crux of a debate between evolutionary biologists Simon Conway Morris, who holds the ring for fate, and Stephen Jay Gould, who argues for chance. But there is another question to pose, relating to the energy behind all of this. For the first time in this book, I will ask, "What is energy?"

The only answer we can give is a short one—we do not know.

This is not a mere gap in our understanding. Energy is something fundamental to everything, yet it remains a marvelous mystery. Energy made *everything* in the universe—it even allowed Michelangelo, through candlelight, to view his progress in the Sistine Chapel in the winter evenings. Those light rays emanating from the candle, like the candle, the building, and Michelangelo themselves, are ultimately bundles of energy. So rather than squeezing God into the gaps, or forcing Him to retreat with each scientific discovery, maybe we could recognize two realms in the universe—one we are capable of understanding, above all through science, and one that we never will understand. Maybe God exists in a realm beyond the human mind. In that case, it may not be surprising that some things appear a little odd, and beyond what we might expect. Maybe one of those odd things is that the creation account of Genesis 1 is a replica of our scientific understanding of the universe and of life. I will explore this idea in the following chapter, where I will attempt to make sense of the Genesis Enigma.

Today, the jellyfish's fragile, transparent body pulsates rhythmically, purposefully driving its searching tentacles through clear tropical waters, occasionally catching a sun's ray and leaving a sparkle in our eye. Although barely visible, it can indeed be seen if we look closely. It gives us a poetic glimpse of its ancestors that swam the Ediacaran seas 570 million years ago. Maybe the first page of the Bible does the same for religion. That opening to Genesis all makes perfect sense when read metaphorically, which may be a key to unlock the remainder of that most illustrious and mysterious book of all.

We have seen how science and the book of Genesis agree, point by point, verse by verse. Science came to its account of the history of the universe and of life through centuries of painstaking research, engulfing the life's work of many vigilant and impartial thinkers, forever fine-tuning the story until it fitted the facts. But how did the writer of Genesis come to his conclusion? The following chapter will carry out the last remaining task—to consider the possible answers to that question.

God

We must now digest the Genesis Enigma and find its place within the science versus religion debate—where evidence for and against God is offered and assessed. Because I have adhered to science yet have raised a new idea for religion, the Genesis Enigma must have a place in this discussion somewhere. This chapter is about searching for its home. It will be important to take a rational, detached, spectator's view of the modern, prejudice-plagued arena where science and religion go head to head.

To sum up, the Genesis Enigma holds that the Bible has, in its opening page, correctly predicted the history of life on earth, with its series of macroevolutionary steps, or fits and starts, from the origin of our solar system to the evolution of birds and mammals. We can be certain that the author of this biblical account would have had no idea of these scientifically established events, covering billions of years—indeed, the final links in this chain have been forged only very recently. The possible explanations for this parallel between the Bible and modern science are clear-cut: Either the writer of the creation account of Genesis 1 was directed by divine intervention, or he made a lucky guess. (Throughout this book I have, for simplicity's sake, referred to a single author of Genesis, but the truth is certainly more complicated and is the subject of considerable debate and research. It is an important point in understanding the whole context of how and when Genesis was written, but separate from the main argument of the Genesis Enigma. For more details on the question of authorship, please see the appendix.)

So far I have painstakingly pointed out that, considering the natural environment and academic atmosphere surrounding the originator and writer of Genesis 1, the lucky guess option supposes a *considerable* stroke of good fortune. Then again, the Bible lacks hard evidence of divine activities to support the alternative explanation. We know only that the *testable* part of the sacred text—the historical accounts of real people, and their chronological order—is definitively true. Now it becomes a matter of which side of the fence you sit on, or where you place your faith. For me a succinct illustration of this division exists on the walls of the Vatican, a stone's throw from Michelangelo's masterpiece that initially inspired the Genesis Enigma. It is a visual representation of the two sides of this debate.

Return to the Vatican

Michelangelo had been treading his scaffolding in the Sistine Chapel for several months when a rival artist arrived in Rome. Elsewhere within the Vatican was a room with a single large window and two doors on opposite walls, forming part of a wide, partitioned corridor. This was to become a library and private office for Pope Julius II—the Stanza della Segnatura (Room of the Signatura)—and in the second half of 1508 he commissioned an artistic renovation. Unimpressed by the existing display of biblical and classical treatments, Julius sought loftier ideas and youthful energy. Championed by the influential Donato Bramante, the pope's architect, Raphael was given the job of providing new frescoes.

Raphael was only eight years Michelangelo's junior, but the chasm between them was a deep one. With his well-groomed, boyish good looks, elegant attire, chivalrous manners and sociable persona, Raphael was the antithesis of the rugged, reclusive genius at work in the Sistine Chapel. Raphael's upbringing in the court where his father had painted had instilled within him a social ease. Although not an intellectual man, he had combined his natural artistic talent with a fervent desire to learn (for example, from Leonardo's experiments with form), an eagerness to please patrons, and an ambition to become one of art's "big three" of the time. In due course he earned that accolade, when in the eighteenth and nineteenth centuries the perfection and grace in Raphael's paintings finally came to be fully recognized and celebrated. But until then, Michelangelo and Leonardo da Vinci reigned.

After his apprenticeship in the central Italian city of Urbino, Raphael became a student of Leonardo's in Florence. Naturally, this meant that Michelangelo, who disliked Leonardo, would also dislike his protégé, although the two managed to work simultaneously in the Vatican with little verbal sparring. Certainly, Raphael knew the political realities of his work. In terms of covering the bare walls of their respective rooms, Raphael's and Michelangelo's tasks differed. While each was to execute the commission in his own unique style, Michelangelo had demanded his own choice of subject while Raphael had (less perilously) accepted the ideas of Julius II.

As described in the introduction to this book, Michelangelo chose to decorate the vault of the Sistine Chapel with the evocation of divine revelation, beginning with the creation account of Genesis 1. Raphael was instructed to follow this with a depiction of the presence of revelation in the actual world. He was to show how real people could be reached by the divine, through either active or passive communication; but however it happened, there should be no doubt that they clearly understood what was being disclosed to them.

Julius II espoused a reasoned, humane approach to Christian life, and he wanted this expressed not with theological or philosophical reasoning, but with the figurative means of art. In the Renaissance, art had become the new instrument with which to investigate the complexity of reality in order to reveal the influence of the divine. This new perspective on God had to be reunited with the spiritual particulars of faith, without losing anything of the intellectual ground that had been gained. The task of art was to symbolize meanings that lay beyond the limited reach of linguistic vocabulary, not only through iconography but through the entire expression of a painting. Michelangelo and Raphael were to translate into imagery the Neoplatonic proposal of a more rational religion—one that sat somewhere between Christian teaching and Greek philosophy.

Both artists succeeded in the task, each expressing in a realistic image the indescribable divine, while abandoning the pure abstraction of the medieval symbol. They had expressed the new religious thought in their art. In one of Raphael's paintings, the scholarly speculation depicted has particular relevance to the science versus religion debate today.

After completing his first fresco in the Stanza della Segnatura, the *Disputation on the Sacrament*, between the years 1510 and 1511 Raphael painted the *School of Athens* on the opposite wall. The *Disputation* represents supernatural truth, where the church spans both heaven and earth—we see here Christ enthroned in the painting's center, flanked by the saints of the church, all seated above their earthly counterparts, the bishops and priests of the Catholic Church. The *School of Athens* represents rational truth, or the truth acquired through reason. These two paintings embody the two extremes in

Christian views today. The *Disputation* could represent the creationists, with their supernatural belief that every phrase in the Bible is literally true as it is written, and a hint of their rigid refusal to entertain alternative arguments. Rational Christians, on the other hand, can claim the *School of Athens* as their portrayal.

Here, under a colossal, semicircular Roman arch, muse the students and scholars of an ancient, fictitious university, sprawled over and standing upon a set of four grand marble steps, the uppermost step meeting a corridor that passes through three triumphant arches, with intermittent swatches of blue sky. There's the sense of an operatic stage—perhaps (and I don't speak as an art critic) a set from a Roman *Aida*. Hardly a third of the robe-clad cast were Athenians, and overall they derive from a range of centuries. They are captured in mid-thought, keenly teaching or feverishly learning. The foreground, shaded by the giant curtainlike arch, is in darkness. The scholars are in the light.

Determining who's who in the painting is not all that easy, particularly since the portraits of fellow artists included representations of some individuals known only to Raphael. But many identities are obvious.

The mathematical sciences are represented on the marbled floor in front of the steps—on the left, music and arithmetic with the Pythagoreans; on the right, geometry and astronomy. The steps in the center express the higher degrees of knowledge successively reached with these sciences.

At the center of the *School of Athens*, to which perspective draws the eye, are two figures—Plato on the left, Aristotle, his former student, on the right. Both figures hold copies of their books (in modern, bound form) in their left hands while gesturing with their right. Plato holds *Timaeus*, Aristotle his *Nicomachean Ethics*. Plato is depicted as old, gray, wise-looking, and barefoot. His heels are lifted slightly from the ground, as if he were poised to ascend to the heavens. By contrast Aristotle, slightly ahead of Plato, is in mature manhood, handsome and well dressed in a gold-trimmed robe, and appears to be attracting the attention of the younger men around him. With his right hand Plato points upward, vertically, toward the heavens, while Aristotle stretches out his right hand horizontally, toward the viewer, with his palm down. He initiates a powerful flow of space toward the viewer. There is, of course, a reason for such imagery.

Plato's purple and red garments symbolize contemplation and a contained will. That Plato's partly obscured feet barely touch the ground as he walks suggests a kind of otherworldly presence.

Aristotle, who is welcoming Plato into the School of Athens, wears a blue robe symbolizing cerebral activity. Not only are Aristotle's feet deliberately noticeable, they are also planted firmly on the ground, departing from the direction of Plato. Aristotle and Plato both believed in an element of the divine, but clearly differed in the degree of science to be found in their philosophies. Aristotle, while still immersed in an academic atmosphere of unquestionable acceptance of the divine, hints that we should not overlook nature as we seek an explanation for the universe. Though both Aristotle and Plato believed in a God of some kind, Aristotle was hesitant to accept ideas as fact unless they had withstood some form of testing. He applied reins to the Platonic idea that mathematical *theory* held all the answers.

The writings of Plato and Aristotle have fundamentally shaped most of Western thinking. Plato had been influenced by his teacher Socrates, his travels in Egypt, Italy, and Syracuse (in Sicily), and the students of Pythagoras. Since they are relevant to this book (as will become evident as this chapter progresses), I will list a few of his central principles.

Plato believed that the world, as it appears to our senses, is in some way defective and filled with error, but that there is a more real and perfect domain, populated by entities or abstract objects (called "forms" or "ideas") that are everlasting and unchanging and form a framework for the structure of our world. Among the most important of these entities are goodness, beauty, equality, bigness, likeness, unity, being, sameness, difference, change, and changelessness.

At the heart of Plato's philosophy is the distinction between the many observable objects that, for example, appear beautiful, and the one object that is what beauty really is—the essence of beauty. The observable objects take their characteristics from this pure form.

Plato set out in his work to demonstrate the immortality of the human soul. His *Timaeus* contains cosmological speculations, setting out his belief that the earth lay at the center of the universe, and the stars lay on a sphere surrounding the whole system.

In contrast, Aristotle indicates with his gesture in the *School of Athens* the worldliness—the concreteness—of his contributions to philosophy. Aristotle was the earliest natural historian whose work has survived in some detail. His most striking work concerns the sea life of the Greek island of Lesbos—for instance, he separated aquatic mammals from fish, and his description of the male's fertilizing arm in octopus and squid was about two thousand years ahead of its time. He was not entirely convinced by Plato's cosmological conclusions.

In other ways Aristotle shared views with Plato. Aristotle was beginning to suggest a sense of self-organization or natural explanations in the animal kingdom, departing somewhat from a strict doctrine that God created everything. Yet at the same time he shared Plato's belief in God. In fact the *School of Athens* can be divided into these two schools of thought—various degrees of Plato's philosophy on the left and Aristotle's on the right. As I mentioned, today these views could all be considered Christian, and varying views on the influence of God can be found within the Christian sector of the science versus religion arena. An exception, however, exists on the far left of the scene depicted in the *School of Athens*, in a space of his own removed from and overlooked by the other characters. This is the figure of Epicurus.

Epicurus (341–270 BCE) viewed philosophy as a means to attain a happy, tranquil existence, in the absence of pain and fear, by living a self-sufficient life surrounded by friends. He taught that death is the end of the body and the soul and should therefore not be feared, that the gods do not reward or punish humans, that the universe is infinite and eternal, and that events in the world are ultimately based on the motions and interactions of atoms (yes, atoms—the concept had just been born) moving in empty space. He denied the popular concept that everything in the universe was predetermined, or was bound by fate. Certainly, Epicurus departed from other Greek thinkers of the time. Remove the sentimentally used word "God" or "gods" from his equations, and Epicurus becomes an atheist. Certainly the followers of Epicurus and the Sceptics are missing from the painting, since they do not fit the Platonic scheme that guides and inspires Raphael—bluntly, they omit God as the Vatican would recognize Him.

Almost a thousand years of thought and belief are covered within Raphael's *School of Athens*, symbolically gathered within an everlasting here and now. In fact the painting remains very relevant to the present. In terms of innate brainpower, we're no wiser now than Aristotle, Plato, and Epicurus were. But successive generations of scientific experiments have led us away from and beyond belief in Aristotle's four elements of earth, air, fire, and water. Physicists have progressed so far that they are confident of the history of the universe back to the first thousandth of a second. Likewise, biologists today are certain that all living things can be traced back to some universal common ancestor that lived around 3,600 million years ago. Science certainly progresses. But what happened during the first *millionth* of a second, and how the first living organism emerged from a muddle of organic compounds, remain mysteries. Maybe the thoughts captured in the *School of Athens* are as useful to solving *these* problems as anything science has uncovered since. What I mean is that where science *can* progress, then it does; but where it does not, then maybe it *cannot*. In other words, there may be a realm of the universe that science can explain, and a realm that it cannot. Our understanding of the realm that science cannot explain may remain static, unchanged since the beginnings of scholarly societies. It does seem that Plato's "beauty," like consciousness, was just as well understood in ancient Greece as it is today. On that note, is it right that Darwin's progress made religion an irrational choice altogether?

Atheism

On July 1, 1858, papers by Charles Darwin and Alfred Russel Wallace read at the Linnaean Society in London made the first public statement of their theory of evolution by natural selection. The following year Darwin published *On the Origin of Species*. The world was both shocked and divided. A first reaction was that this was the end for God. This reaction has lasted. The Western world remains strongly polarized with respect to God. The "defense lawyers" on both sides argue their cases with some bias.

After writing much of this book, I played a role as science adviser for the Natural History Museum's evolution gallery. My job was to work with a design company and in particular to hunt down recent advances in the field of evolutionary study—of which there have been so many that the gallery is set to become the largest in the museum. That single book, *On the Origin of Species*, clearly triggered the greatest revolution in all biology, if not all science.

In my role in the development of this gallery I was forced to meet one word head-on—"theory." It is a word no longer apt in this context, for the weight of evidence for evolution has become crushing. The "theory of evolution" has become "evolution, the fact." This is important to religion.

If evolution is considered merely a theory, then it is plunged unfairly into debate. By definition, all "theories" allegedly solving the same problem must begin with equal standing—in which case, anyone could invent a theory to account for the diversity of life (that all species came from outer space, for instance) and be given equal standing on a stage with Darwin and Wallace. Then a cunning creationist would argue that Darwin and Wallace's theory is just one of many that account for life on earth, fraudulently detracting from the achievements of these great scientists and at the same time misleading the public. As I said, evolution is a fact. I will come to creationism later, but first let's look at what happens if we *extrapolate* evolution beyond, perhaps, the original purpose or intentions of its first exponents.

Although Darwin accepted that the truth he had uncovered placed religion in doubt—such doubt, indeed, was later to cause the Christian captain of the *Beagle*, Robert Fitzroy, to commit suicide—he went nowhere near as far as the modern atheists in tearing God, as a supernatural creator, to shreds. The atheists' world is a dark one. We live until we are eighty and that's it. We are but a few grams of salts and several pints of water. The universe is made up only of matter as can be explained by our laws of physics. There's no soul, everlasting spirit, or higher force to call upon, and concepts such as fate and destiny are illusions. But that's not what Darwin was about. In this sense we should leave him out of the debate.

It is noticeable that atheist academics obstinately quote Darwin. Darwin has, indeed, become their patron. But we have seen that *both* Darwin *and* Wallace drew the conclusion of natural selection independently. So why do atheists not study Wallace's text with equal vigor? Probably because Wallace was spiritually inclined, and remained so throughout his life. Also interesting is that a good number of Nobel laureates believe in God. Although atheism has good public relations, it is far from the overwhelming rule for scientists today.

Certainly, I do not buy into the atheists' claim that the physical stuff in the universe—that explicable by science—provides richness in our lives that can compare with biblical religion. That we are representatives of a long, long line of ancestors that have made it through life's vast struggle for existence, or that we are connected (via evolution) to every other organism on the planet, is supposed to offer some kind of purpose, meaning, or warmth; something to cling to in what would otherwise be complete emptiness and nothingness, maybe. At its extreme, science is portrayed as more wonderful than the concept of God. Darwin's use of the word "grandeur" is often invoked when referring to nature, and everything, everywhere on earth, becomes "breathtaking"—a profound work of art. But let's face it, the elegance and beauty in hummingbirds and hills will not provide the solace fashioned by a belief in "life" after death, nor will it provide comfort in people's individual lives. The word "individual" here is perhaps key to the disillusionment with the idea that science can ever replace religion. Science can offer some wonder to us as a species, but maybe only religion can provide guidance and purposefulness to the individual passing through his or her daily routine—particularly when that routine is lived out in a concrete city.

I have stated that evolution is a fact. I see no problems whatsoever with the process of evolution. But I have encountered questions about life on earth that should have an equivalent scientific answer, yet to which no such answer seems forthcoming. For instance, why do living things reproduce? This is easy to answer where higher animals are concerned, where juveniles eventually look after their parents. But where the offspring disappear over the horizon or into the depths, the case is less straightforward.

If asked what would happen to the *first* living organism to evolve, I would not predict "reproduction." What benefit to the first individual could be gained from reproducing? If anything, reproduction requires an energy sacrifice. Science tells us what happens during the process of reproduction, but I can't find an explanation for why it happens at all. And then, given that individuals do reproduce, why do they speciate?

Here, as is most often the case, science *can* provide an explanation. Speciation happens because genetic mutations occur—errors during the process of gene duplication. But *errors*. Plato alluded to our defective senses, making a point of errors. Errors are necessary for the variety of life that we have—for evolution, as I have just mentioned. Geneticists take for granted that "a mutation has occurred" here or there on a genome, but such error, at the fundamental level, is without an explanation.

Maybe the origin of reproduction could be considered an error. But why should errors occur? Again, if I were to guess how the first organism reproduced, I would suggest "accurately." A parent should give rise to an exact replica of itself. This, surely, is parsimony, the "less is better" concept that underpins science and is applied in Occam's razor (the idea that plurality ought not to be posited without necessity). It follows the laws of physics. But then, there are many other cases of errors in the natural world. The decay of an atom involves the error of energy loss, and chaos (as it is found in nature) occurs when the organization of systems breaks down. Maybe errors on the molecular scale filter through to the macro level of human behavior. Are the same errors, I wonder, accountable for human tragedies?

Errors provide evidence against the creation—God's system must be error-free, and yet, as we see in our world of science, it is not. But maybe, since they represent a departure from parsimony, errors provide a challenge to science, too. A possible explanation for errors is that they are part of a purposeful system, a tool by which to achieve evolution, for instance. Could the parsimonious scenario be that God created the original universe—the energy or matter—along with the necessary potential for errors, and from here everything self-assembled in line with our scientific explanations?

It is interesting that the writer who penned Genesis 1 depicted God in a form unlike humans or anything we can conceive of, but also as a God who could sometimes appear harsh. This is consistent with a God that allowed his building blocks to self-assemble, with the added potential for error. In which case, disease and other adverse occurrences become explicable, averting the question of how a good God could let such things happen.

At least, these are *my* observations and thoughts as an onlooker of the science versus religion debate. I am trying my best to be impartial, attempting to pick the virtues from each side of the argument. Again, I am a practicing scientist and just as I will seize upon anything unreasonable in a research paper or grant proposal that I am asked to review, I will weed out anything irrational in this subject of science versus religion, too. Soon it will be simple for me to declare creationism dead on the ground, but here I am just beginning to find incompleteness (not inadequacy, but *insufficiency*) in *science* as an explanation for the universe, too. I am also warming to the idea that we are better off, as individuals, living with religion than without it. Needless to say I do not refer to religious extremism, which is *not* an inevitable consequence of the Sunday church service or the Bible that is digested rationally.

On the other side of the science versus religion arena sits a cause of far greater trepidation—the creationists, or those who take the Bible's words literally.

Creationism

Choosing the themes to be included in the Natural History Museum's evolution gallery was fun, but wading through the public's perception of what the gallery should contain became a more serious experience. Many felt that the creation—and its surrogate, intelligent design, which I will come to—should be offered as an alternative to evolution.

Indeed, a recent poll revealed that 45 percent of Americans don't believe in evolution. Thirty-nine percent of Britons believe in creationism or intelligent design by God as a better explanation than evolution. Very large minorities in both countries believe that these "theories" should be taught in schools. You see my problem with "theory." This result is dangerous, and here's why.

Fact: Humans and chimpanzees share 99 percent of their genes. This statement would shake the earth beneath those already troubled by the very words "Darwin" and "evolution." But this really is fact, no bones about it. Unlike Darwin and Wallace, we are able to sample the DNA of a human and a chimp and come to this conclusion. I could do it, and you the reader could do it, and we would find the same result, time after time.

Similarly, examples of the same "Burgess Shale" animals discovered by Charles Doolittle Walcott have emerged as fossils in rocks of very similar age but from very different parts of the world. The rocks in this case were independently tested and the results gave us ages in the range of 508 to 519 million years old. It's that precise and accurate. Simply, these animals *were* living at that period in time.

This is what makes evolution powerful: It is testable. Religion is not. Yet clearly many people trust that the Bible's every word is true, in which case the earth is just six thousand years old. They are perfectly entitled to believe this, of course—faith is faith—but in doing so they are turning their back on evolution, which we know to be factual. Crank philosophy results. These "creationists," or fundamentalists, become harmful to science, however, when they concoct theories that reinterpret the biological evidence, and then advocate the teaching of these in schools. In my opinion, intelligent design cannot be better described than as a concocted theory. According to intelligent design, there is design inherent in all species, so it follows that there must be a designer. But it certainly does not follow.

Try as I do to be fair and open-minded, I can't help but be struck by the flawed logic and forced theories found in the case studies of intelligent design. Deep under the skin of each example evolution boils vehemently, but is suppressed and silenced beneath the idea that complex and sophisticated structures in nature must have been built in complete form from the beginning of their history. There is no room for incremental steps, from simple to intricate, in intelligent design theory. It is argued that the theoretical intermediate steps could not function in their own right. But when we look at those transitional stages in the case of the eye, as shown in chapter 6 for instance, we find that each *can* function in its own right, with each stage providing an advantage over its predecessor. I could build a model of each step, and the data gathered would multiply as the steps increase in complexity. In fact the idea of intelligent design is considered ridiculous by almost the entire scientific community. Accordingly, I would ask advocates of intelligent design to redirect *their* intelligence and efforts toward developing an alternative religious stance. Not only is their work detrimental to science, it is also unnecessary to Christianity.

There are other instances of plainly erroneous thinking that are also damaging the case for religion. "Flood geologists" have proposed many theories to reconcile the sequence of fossils evident in the fossil column with the literal account of Noah's flood in the Bible. Some, faced with the self-appointed task of explaining why different types of fossils occur at different depths from the earth's outer surface (in rocks of different ages), have proposed that, because organisms have different shapes and densities, they also have different buoyancies, and this determined the sequence in which their remains settled to the bottom of the flood waters. Other theories hinge on the different behaviors of different animals or their different positions in the water column, which affect where they finish up in the burial process. Clearly, considering the literal timeframe indicated in the Old Testament (in units of days), the fossils all had to have taken their places in the fossil column quickly. As described earlier in this book, modern geology demonstrates with testable evidence that the fossil column formed over millions of years, by means of mostly slow-acting forces that can be seen in operation today. Then, of course, the theory of natural selection and the molecular clock reveal that the animals and plants represented in the fossil column are separated by millions of years of gradual change and evolutionary branching, so that those appearing deepest in the column are the more ancient species, not the animals that lived deepest in the water.

I agree that no one has the right to tell anyone what they should believe in—but come on, all you advocates of the Bible's every word: If you *believe* simply because your faith tells you so, then just accept that *faith* is your only argument. This would be far more dignified than childishly mangling the facts and concocting a dishonorable theory into which they can be stuffed. It's no wonder that scientists cannot take such theories seriously; and, worse than that, the scientific world will begin to label religion as unintelligent and unreasonable. That's *religion* in general—even the more rational approaches to understanding faith and the divine, which I hope would include the Genesis Enigma, will be seen as tarnished. But there are far graver reasons why we must not overlook evolution.

Consider medicine. Microbes also, unfortunately for us, undergo evolution. A dose of an antibiotic is met by genetic mutations that eventually make the microbial targets resistant. Virulence evolves. What is true for microbes and antibiotics is also true for crop pests and pesticides. The first stage in attempts to counter virulence is to understand the evolution of these organisms.

Evolution tells us that the origin of the human immunodeficiency virus (HIV) was precipitated by a jump across the primate species barrier. It enables us to predict the imminent arrival of avian flu and the mutations most likely to be responsible for that evolutionary leap from birds to humans. There are now countless examples of where an understanding of evolution has come to the aid of medicine. Many of these center on our far removal from the environment in which we evolved, and to which we are adapted. Original, hunter-gatherer humans did not eat a diet high in fat and sugar—today's obesity, hypertension, and tooth decay attest to that. Inevitably, the race is on to alter our genes to provide us with new bodies that *can* cope with contemporary living.

How dangerous to ignore, let alone deny, evolution.

Certainly, in the twenty-first century it is unreasonable to accept every statement in the Bible as the literal truth. In Victorian times, the King James version of the Old Testament was regarded as having been dictated, from the first word to the last, to its transcribers by the Holy Ghost. It was affirmed that not merely the doctrine but the actual language was divine, incapable of error and superior to all the frailties that accompany human literature. The Victorians considered any Bible that differed from the King James version to be imperfect.

Making trouble for himself, the French priest and philosophical and political writer Félicité Lamennais challenged such conservative views of the church with his book *The Words of a Believer* (*Paroles d'un croyant*, 1834). In 1906 a series of drawings by the Swiss artist Carlos Schwabe were commissioned to accompany the original manuscript, and these caused further controversy by drawing attention to the absurdness of the literal interpretation of the Bible. One example—"Notre Père avec son baton"—shows God in human form, in an image taken directly, literally and crudely from Genesis: a worldly old man, although a little weary-looking, deserving of a rest. He has just created the universe, after all. This depiction instantly becomes a mythological image, then even a parody. It's hardly surprising that intelligent academics have a problem with it. But soon after Carlos Schwabe's graphic exposition, the lessons were heeded. A charter of critical freedom was drawn up by the Catholic Church and Moses' rod no longer turned into a *real* zoological serpent. So how much ground toward making religion rational has been made since then?

Middle Ground

Acceptance of evolution within the church is certainly mounting. In 2005 the hierarchy of the Roman Catholic Church admitted that some parts of the Bible are not actually true. They drew attention to the fact that the Bible expresses God's word in human language, leaving room for modern interpretation of the holy words. Also in

2005, the Catholic Church of Scotland published a new guide to the Bible that explained that some passages are not literally true. Worshippers were warned not to expect "total accuracy" from the Bible. The Garden of Eden and the creation of a woman from Adam's rib were dismissed as "symbolic language." The guide, on the other hand, does accept as true God's revelation to Moses.

Well, all was progressing nicely until, in the following year, Pope Benedict XVI sacked his chief astronomer for speaking out in the same vein. Fortunately, some sectors of the church are more liberal. Recently the Archbishop of Westminster went so far as to call for respect for atheists and agnostics. Maybe his idea, that there was a "hidden God" active in everyone's life, was a step in the right direction, even though it does not tell us a lot. For me, we could do worse than explore this concept of a hidden God, but without trespassing on the facts of evolution, and science in general.

With this in mind, when we read anything from the Old Testament, we have to ask whether it is providing information or acting as a symbol. If the words are symbolic, we must then consider what implications they had in their time.

The conclusion I am approaching is that accepting evolution as true does not make me an atheist. The creationists are out for the count, but science has not proved all-conquering, either. So where next in my search for rationality in religion?

There has emerged from Oxford University's Green Templeton College, fueled by the enthusiasm and wealth of knowledge of mathematician Professor John Lennox (introduced in the preface), a group of scientists who are exploring the middle ground between creationism and atheism. A practicing scientist gains a feel for the way science works and progresses, and also for its likely expansion in the future. One or two problems may exist, nevertheless, where we cannot foresee science ever providing a solution. These are cases that simply are not suited to the methodologies of science. I have in mind at this point the origin of the universe, or of the energy that resulted in the universe, or what lies beyond the universe. The question, "Why is there something rather than nothing?" often comes up. Hold on; let's keep our feet on the ground.

First of all, we should recognize how not to rationalize religion. Many scientific authors have attempted to find some middle ground by blurring the argument. They have changed the definition of God to something less humanistic, paternal, or personal—"God is nature," for instance. Well, many atheists have taken responsibility here and set the record straight. They have given the harsh lesson that you must believe in God as portrayed in the Bible—a God to be worshipped—otherwise "He" becomes spread so thin as to be useless. To believe in God is to believe that the universe has a purpose. Indeed, there exists another course to rational faith. If one lays the facts of the universe on the table, there is actually something left after science has accounted for all it can account for, or is likely to account for in the future considering the way that it works and progresses. Such a view requires a complete lack of any bias, including the atheist-style faith that science will eventually account for everything. I suppose I am considering God as something that created energy from nothing in the very beginning—the entity behind the cosmological Big Bang as described in chapter 2—and perhaps as a force with a *purpose*, dissimilar to the forces we measure with scientific equipment.

When I suggested a way forward at Green Templeton College, I was met not with the joviality I expected, but rather with a deafening hush. I could feel the words being drawn from my mouth. I began with the obvious, that science is rock solid. You can't tinker with that. But I questioned whether religion existed in a totally separate ether, whether there's a place for religion beyond the scientific realm. Let's accept science and get on with finding God within these new boundaries, I argued. I was taken seriously. This was because my audience had drawn similar conclusions themselves.

And not only my audience in Oxford. Many post-Darwinian thinkers have found God within the confines of the evidence of evolution and far from the claptrap of creationism. I will not make a list of those in this category, as a way of boasting, "Look how many support this!" Rather, my purpose is to examine *what* some of those people had to say on the matter. I will begin with Clive Staples Lewis.

The Thoughts of C. S. Lewis

C. S. Lewis wrote much about a law of human nature, or "Moral Law." This is something within us that compels us to goodness. It is unlike instinct, which is inherited from our parents and may impel us in different directions. Instinct takes hold when we witness an assault, for instance. One desire, derived from our herd instinct, is to help the victim. Another desire, derived from our instinct for self-preservation, is to protect ourselves and to stay clear. But, separately from our instincts, the Moral Law coerces us to do the right thing. This is something we feel inside us. It tells us to follow the impulse to help, and suppress the impulse not to assist. The Moral Law can explain why prisoners of war were kept alive and other inexplicable acts of compassion. It is why humans have a curious idea that they ought to behave in a certain way, and cannot really get rid of it. You may want to stay safe rather than save a man who is drowning, but the Moral Law tells you to help him all the same.

The Moral Law is something innate. It is not something we learn, and it is something we cannot discard. Even if we do not obey the Moral Law—if we do not do the right thing—it will remain as a voice inside our heads, conveying a feeling of guilt.

Some of this behavior could be attributed to altruism—the selfless concern for the welfare of others or, in evolutionary study, the behavior of an individual that, while decreasing the fitness of that individual, increases the fitness of another. I give a definition that refers to evolution because altruism does indeed have an evolutionary explanation. For instance, in a group of Australian Aboriginals, or Siberian Inuits, altruism toward other group members was found to improve the overall fitness of the group. Since the group becomes fitter with altruistic behavior, altruism can be explained as a characteristic that has evolved. But the Moral Law also includes instances of behavior that provide no conceivable positive value to either an individual or a species. If we fail to own up to receiving too much change at a supermarket, we walk away feeling an ache in our conscience, yet negligible harm has been done to any individual—who cares about depriving a multinational conglomerate of a few pennies?—and there has been negligible gain to ourselves. No, humans appear to possess something additional to the altruism found in other species. This add-on quality lies at the heart of the Moral Law.

C. S. Lewis pointed out that the moral teachings of, say, the ancient Egyptians, Babylonians, Hindus, Chinese, Greeks, and Romans are very similar to one another and to our own. That's because, Lewis argued, the Moral Law exists in all humans all over the planet. But again, although all humans are forced to believe in a real right and wrong, and know that they should behave in a certain way, they do not in fact behave in this way. They know the Moral Law, and they break it. C. S. Lewis believed that the Moral Law and the fact that we break it are the foundation of all clear thinking about our universe and ourselves.

C. S. Lewis went further to suggest that what lies behind the universe is more like a mind than it is like anything else we know. Here,

God is defined not as the anthropomorphic form figured at the beginning of this chapter, but as so far removed from our perceptions or imagination that we cannot begin to reconstruct or ever conceive of "His" form or kind. The first-century Jewish philosopher Philo Judaeus (Philo of Alexandria) was credited as the originator of such thought, considering God almost totally unknowable and beyond comprehension of our ideas. But maybe the writer of the Pentateuch also held this view several centuries earlier, and if—as I will suggest later—the solution to the Genesis Enigma is that this writer received divine intervention, then, through his heavenly experience, he knew something we don't. Whatever the case, ground is made here toward a logical religion. This view of God is probably critical to the science versus religion debate, particularly when paired with knowledge of how we perceive our world, as I will explain.

Our senses are extremely limited, but this is something very difficult to understand. And this must present a problem when it comes to deciphering God. We know our universe from our senses alone (well, in addition to some scientific detector equipment). It follows that *since our senses are finite, we cannot ever know everything about our universe*. This is the all-important fact which, I feel, is often overlooked by atheists. What we know about our universe may be merely the tip of the iceberg. I will talk more about this subject later in this chapter. But I want to introduce the point here because it is what C. S. Lewis alluded to in that remark about God being more like a mind than anything else we know. We must accept that we will never understand the form of God, or what "He" is. This makes the science versus religion debate less clear-cut. It becomes less simple than rational people choosing atheism and irrational people, in need of some meaning in their lives, opting for religion, as intellectual atheists allege.

If we do not know what one faction of the debate is representing then how, indeed, can we engage in debate at all? We can all accept the facts revealed by science—evolution, gravity, the speed of light, and all the evidence set out over the course of this book. But whether you believe in God is another issue altogether. Science will provide no evidence of the divine. Even if it did, we would be unable to interpret it, since we do not know the form of God. The human mind and body are useless here. We are limited both in our perception and in interpreting the stimuli that we do perceive. Now we are being asked to sense and interpret things beyond these limits. So does this make atheism a *faith* that God does not exist because science will reveal an exclusively material universe? Does science hold the faith that there really is nothing we can't interpret? On the other hand, this could be construed as an argument for belief by faith alone. Atheism is being attacked because it is unable to prove a negative. Nevertheless, a sensible conclusion may be that, considering the way our senses and brains work, the potential for God's existence exists.

Working within this framework, if there *is* a God then it would not be surprising that some things are encountered that do not make sense, or cannot be explained with science. The Moral Law falls into this category, along with why there is a universe at all, how it formed, and what it formed from. I will come to these subjects soon. But here it is interesting that, if God exists, there may well be mysterious elements in our universe, and the idea that there are should not be surprising. These elements may become evidence for God, or at least evidence that there is something beyond the scientifically explicable universe.

An Inconvenience to Science

Personally, I foresee a scientific explanation for many mysteries, but not for the origin of everything—another mysterious element. Here I can agree with Arthur Conan Doyle, who wrote in 1917:

> Napoleon's question . . . "Who was it, gentleman, who made these stars?" has never been answered. To say that the universe was made of immutable laws only put the question one degree back as to who made the laws. I did not, of course, believe in an anthropomorphic God, but I believed then, as I believe now, in an intelligent Force behind all the operations of Nature—a force so infinitely complex and great that my finite brain could go no further than its existence.

Whatever theory is postulated in science, a prerequisite is required. The Big Bang origin of the universe required energy. Newton told us that mass (rocks, for instance) and energy are interchangeable, but also that energy can be neither created nor destroyed. As I mentioned earlier, energy is interesting, although we still don't understand what it is. Many have suggested that from the initial formation of energy everything else in the universe could have self-assembled. Certainly in the biological world, energy is the primary currency used by natural selection. But something must have created energy, presumably at a *beginning*.

Something else we can learn from science on that subject is that, although infinity is a useful concept to employ in theoretical calculations, infinity does not exist in the real world. One could imagine that an object traveling on a circular path, with no beginning or end, would continue its journey infinitely, but we never find this in the real world. From electrons "orbiting" atomic nuclei to planets circling suns, everything runs out of steam at some point. Returning to Raphael's *School of Athens*, Plato would have predicted an infinite universe, while Aristotle would have pointed out that although the theory is correct, infinity is actually absent from nature. Without infinity, it follows that everything in the universe is finite, and so everything has a *beginning* as well as an end.

As I have mentioned already, we could argue that science will answer the question of "the origin," although it has yet to progress that far. Maybe large hadron colliders (particle accelerators) will unearth some evidence. But is this not itself a belief—in science? Again, although science will probably answer most questions remaining, there are no signs to suggest it will be useful in respect of the origin. The Greek philosopher Epicurus maintained, "It is best to keep an open mind in the absence of decisive verification." Maybe there *is* room for God in a rational arena today.

The latter part of the passage from Conan Doyle quoted above brings me to my second argument for openness to the idea of God. I research the subject of vision, and even with our eyes we do not see everything there is to see. Bees and birds communicate within the ultraviolet portion of sunlight, a part of the spectrum humans don't see. I've always known about this, but only after many years of studying this fact has the message finally sunk in—we live in a virtual reality world.

A good way to picture this is to consider that if we did not have eyes, then all we know from vision would be lacking. If we had an additional sense, on the other hand, then we would know more about the world. Go back 550 million years, and animals did not possess eyes. The stimulus for vision—light—did indeed exist. Light reflected from the bodies of neighbors was projected on to the heads of these ancient animals. But they made no use of it. Potential images counted for nothing. Color remained an invention of the future. Nevertheless, those ancient animals would have felt that they knew everything about their surroundings. We feel that we do today, but I can say for sure that we don't know about the world in ultraviolet, nor that in infrared. How far this list of the unperceived can be extended may never be known.

The atheists contend that we are just some salts and a bucketful of water, but let's look at what those ingredients make. Contained within a skin and supported by a skeleton, we have a brain that collects information from a nerve network, which services our major sensory organs. These include our eyes, ears, nose, tongue, and touch and temperature sensors in our skin and other body parts. From this information the brain makes a decision on how best to react, sending electrical signals to our muscles through a separate network of nerves.

The information flowing from our sensory organs is derived from the outside world into which our body is plunged. Within our sensory organs are cells specialized to respond to certain signals. For instance, retinal cells in the eye contain different molecules that become altered by light rays of different wavelengths between 400 and 700 nanometers, and the inner ear contains hair cells that are moved by sound waves in the range of frequencies 20 to 20,000 Hertz. But that's it—that's our *only* means of contact with the outside world. If something lies beyond our range of detection, we will know nothing of it, and we will not react to it. Our brain will not receive information on it, and our muscles will not be told to react to it. If there is an entirely different form of information in our world then, again, it will escape our perception.

Furthermore, our sensory system distorts the information that we *do* collect. There is no such thing as color in the real world, for instance. Color is made in the mind. In the outside world there are light rays with different wavelengths, but they have no color. When the information on these from our eyes arrives at our brain, the different wavelengths are assigned different colors. This is easy to discover, but hard to come to terms with.

Similarly, we look at a rock and perceive a solid block of material. Actually we are viewing swarms of subatomic particles separated by lots of empty space. In fact, as we've noted before, over 99 percent of the rock is empty space. Our limited senses and processing center insist that the truth is a completely solid rock with no empty space. This tells us something of our almost crazy perception of the world we live in. It also makes a mockery of truth. In chapter 1 I mentioned that, for the sake of our investigation, we could consider truth as what really happened. This is now exposed as a simplistic definition, for how do we know "what really happened"? Our whole means of perception and understanding of what is happening around us is suddenly beginning to look hopelessly inadequate.

Let's think further about this. Our brain has a maximum size, which means it can do limited processing work. It is fed a portion of the information in the universe—the most that it can accept and process, or do something useful with. So do we gather and understand *half* of what there is to know in the universe? A tenth? A thousandth? Maybe it is not surprising that mysterious elements exist.

Returning to the theme of vision, we could question whether evolution alone can explain beauty. Some rocks and blocks of material come with beauty, others don't. Yet ultimately all are subatomic particles and space. Some, though, clearly trigger extra activity in our brains. We do not know why.

If we were to recreate life on earth using only evolution—survival of the fittest, and so on—would we expect to develop a sense of beauty in our environment? Would we expect to sense drama when faced with a soaring, shattering ice cliff in Antarctica or a rhythmic avalanche in the Himalayas? Why should we feel emotional about these sights? These feelings may not arise from natural selection with its underlying principle of efficiency—is energy not lost here?

As to the question of beauty, I once acted as consultant for a cosmetics company, which tried but struggled to find a scientific formula for visually attractive hair. There were some characteristics, such as shine and depth of color, that could be assigned a quantitative value, and by changing the physical characteristics of the hair the attractiveness would increase or decrease according to a formula. But other characteristics, such as the perception of health and warmth, or vitality, were not linked to the structure of the hair by any formula. These characteristics somehow evoked the emotions, but we have no idea how that works. This is beyond our current scientific understanding, and may be touching on a part of our being analogous to the Moral Law—that is, there may be an immaterial explanation.

To balance these thoughts, we should turn to a strictly scientific approach and consider the comment of Francis Crick that "You, your joys and sorrows, your memories and ambitions, your sense of personal identity and free will are, in fact, no more than the behavior of a vast assembly of nerve cells." So back to the "salts and water," then, although this time the message is that it's all in the brain. Still, if the information doesn't get in, it can't be processed.

Well, at least we have established that our perception of the world is not the same as that of a beetle or a bat. Individuals extract as much information from the world around them as their sensory system will allow, but much more remains of which they are entirely unaware. We must accept this. Is it worthwhile considering that, among the things we do not sense, there is something divine? This is an interesting subject to explore, and one we can explore while remaining rational and unbiased. I will admit that the notion of God has always appeared preposterous to me; but, placed in the context of our limited senses and interpretation of sensory data, it may indeed be a plausible notion, because now . . . well, who knows?

An Inconvenience to Atheism

Perhaps the greatest inconvenience to atheists is the fact that the Bible remains the centerpiece of our library, after all these years. It still sits on our tables, inviting our interrogation. It just won't go away.

Also lingering is religious *thought*, ingrained in human minds. Not only has the text persisted, its psychological counterpart—the Bible within our bodies—can be found very much alive today, too. The fertility pioneer Robert Winston has made the interesting point that spirituality—the feeling of something beyond mortal life—was important in human survival during the ice age. In this case, spirituality lifted human morale during a period of great deprivation. Continuing along these lines, many people say simply that atheism just doesn't feel right. The dark world of atheism evokes some form of discord in the brain. Robert Winston continued his reasoning to raise an interesting alternative—that spirituality *does* exist, and that the question is, rather, whether it is God-given or actually evolved because it was needed. Now there's an interesting alternative—that the concept of God *evolved* to raise our emotional state, in which case God is a concept only, and does not actually exist. Certainly, I am beginning to concede that the stronger our belief in God, the higher our morale.

While examining the paintings of Michelangelo and Raphael discussed earlier in this chapter, I met many Romans, from many different walks of life, and chatted with them about a wide range of subjects. From these conversations I began to sense that in Italy, in general, religion is integrated into the common psyche, and that it gives people a real sense that no matter what happens in their world, they themselves will be all right. From baptism, through family upbringing, schooling, and working life, the historic Catholic Church instills within Italian people the sense that God is everywhere, incorporated into everything—just as in the ancient Middle East, as T. E. Lawrence noted of the Bedouin people in his day, there is (almost) no separate word for religion. God is not separate from matter, feelings, or activities, or from academic subjects including science. An atheist could not remove God from the life of many Italians if he tried.

There is only so much any of us can do to change our destiny, or improve our life. Yet in Italy, in general, no obstacle will ever appear insurmountable. The glass is always half full, never half empty (an expression I encountered often in Rome). I could not help but notice a distinct spring in the Italian step. The employment prospects are not always rosy, but the complete faith that God is there for the individual lessens fear about the financial future. In my opinion, science can't achieve this.

Faced with these observations, I can't help asking: Is the atheist crusade to destroy religion really a positive step, or even necessary? Probing further into my Italian experience, I'm wondering whether religion is a critical element in the human mind, a link in the *physiological* chain that, when complete, makes for happiness and emotional balance? In this view, God is synchronized—in harmony—with the human mind, in that the divine ether is interfaced with the matter in our bodies. Such an interpretation could account for C. S. Lewis' Moral Law, and could help us to understand why a fully integrated God—a situation where there's no separate word for religion—is better for us. Alternatively, applying the other half of Robert Winston's theory, spirituality may have *evolved* as a link in our psychological chain.

All this discussion, on the religious thoughts of C. S. Lewis and others, has brought us to the point of acknowledging that there are "mysterious elements" of various kinds in the universe, elements without or even beyond scientific explanation. Returning to the reason for this discussion, it is time to question whether it is here, as one of these mysterious elements, that we have found a home for the Genesis Enigma.

Could It Be Real?

What if a mysterious element were contained within the text of the Bible? Of course, I'm not talking about the literal interpretation—creationism—but I'm wondering whether those sacred words were trying to point us toward something, such as the Natural Law. Is the reason why the Bible remains on our tables that it really does carry a divine message, and that this message strikes a chord with the somehow receptive human body?

It has been suggested that in the *School of Athens* Raphael achieved far more than he was capable of—that he did not realize what an important contribution to religion he was generating. From what we know of him, Raphael would have been conscious only of the composition, accuracy of form, and grace in his painting, not of the deeper religious implications. He would have been unaware of how he was bringing into harmony the spirits of antiquity and Christianity. Some viewers sense God's hand in this (and other) artwork, asserting that Raphael has described human cognition above a spiritual threshold.

Returning to the holy text, we must recognize a problem with mere words, in that they cannot accurately represent a picture or event. They do the best they can, but everyone would use different words to describe the same scene. Certainly, it is only through metaphor that we can express in language the sense of an energy. And we know that at the time the Bible was written people wrote in metaphors.

Now, the Genesis Enigma has indeed exposed evidence that the Bible is a metaphorical text—if we treat its first page metaphorically, we find truth. But it remains extraordinary that, against all odds, the writer of the creation account of Genesis 1 got it right in that the skeleton of his exposition was scientifically accurate. The image in this writer's mind of how our planet and life came to be must have seemed so far off contemporary knowledge and experience that he was unable to dress it in human clothes and could only present it unaltered, as if it had been dictated to him. Yet the words he wrote later emerged as an exact match with modern science. This all does appear as a "mysterious element" in the vein of those I have discussed above. If my inference is right, then the writer of Genesis 1, or rather the announcer of the story—Moses—surely must have received divine intervention. That is to say, he must have been spoken to by God. The image of life's and the universe's origins formed in the mind of the writer of Genesis 1 or Moses must have been placed there by God. I would argue that *the Genesis Enigma, under this line of reasoning, becomes evidence for God*. Otherwise, Moses, or the writer of Genesis 1, must have understood that the universe formed first, then the seas appeared on earth, and that life forms were photosynthetic. Further, he must have realized that an eye evolved in an early animal in the geological past, which triggered the evolution of all the major groups of animals that exist today. Further still, he must have felt that this all occurred in the sea, before animals moved onto land, and only when they did move out of the water did mammals and birds evolve. This would appear an incredibly lucky guess. The alternative explanation is divine intervention. This interpretation echoes the Catholic Church of Scotland's 2005 guide to the Bible, which, while arguing for a symbolic interpretation of the text, does accept as true God's revelation to Moses.

The paragraph above is my most daring and contentious in this book. I'm acutely, uncomfortably aware of this. Yet I have stuck to the facts, and have tried to be impartial. Still, I am a scientist, and incorporating God into my thinking feels like a quantum leap. But I am at least comfortable with the idea that my acceptance of evolution, and science in general, does not affect, or is not affected by, my feelings about God. Maybe it *is* wise to consider that God exists, in unimaginable form, beyond the material world that can be explained by science. If so, maybe "He" occasionally manifests in the mysterious elements that we cannot otherwise explain. I do feel comfortable to go so far. In the future, my "maybes" could become something more decisive and confident.

Researching this subject has made me wise to out-of-context quotations and loaded questions. If asked, in an academic setting, "Do I really believe in all of this?" then I would feel drawn toward the answer "No." Even "Of course not." In such a scenario, someone is presenting me with the picture I decribed in this chapter, of God in human form sitting on the clouds. To dissuade any rational person from religion, just paint this picture. It does, indeed, adhere to God's name like iron to a magnet, since our first stage in thought involves anthropomorphizing.

Then again, if asked, "Considering that my senses are limited and do not discern the entire world around me, and that my brain distorts the information I do receive, do I accept the possibility of a God in a form that I could never perceive or comprehend?"—well, now my answer would be different. The image of God has become more haze than human. God has become something like the "mind" suggested by C. S. Lewis. Maybe God is a force with a purpose that could assign direction to the energy in the universe (initially created), with the potential for errors that could lead to the necessary change that has occurred on earth. Maybe Erasmus Darwin was right with his idea in the *Temple of Nature* that God is the cause of causes, rather than the cause simply of events. Indeed, the writer of the Pentateuch imaged a God who sets the rules and that's all. In which case, God and evolution can coexist—the conclusion supported by the Genesis Enigma. I don't think that Erasmus's grandson Charles Darwin did anything to change this train of thought.

To finish off, I would like to share a conversation I had on a bus in Boston while writing this chapter. Returning from a scientific conference, I sat next to a distinguished medical researcher, Dr. Lynda Bonewald. Lynda works in Kansas, home of the anti-stem-cell research lobby, but it was only after a week of fascinating scientific exchanges that she oversaw a draft of this chapter and politely revealed her hand.

Lynda is friendly, level-headed, and focused on her work. She flies the flag to support stem-cell research—expending a lot of energy, out of a sense of duty, on what should be an unnecessary job. The religious movement she contests takes the Bible literally and quotes passages that condemn her work. But she knows the importance of that work to humanity, and so must lobby the U.S. government to continue funding it, so that ultimately lives can be saved. We are back to the link between evolution and medicine. But the interesting fact at this point is that Lynda believes in God. Her stance on stem-cell research is based on the tradition of virtue ethics, the style of thinking handed to us by Aristotle. Today, wisdom, prudence, fortitude, and temperance provide the guidelines for the exercise of public responsibilities.

Lynda cannot, despite struggling to do so all her life, understand why she believes in God—she simply has an inexplicable sense that God exists. This reminds me of a related point, that many of us find the atheist mantra—that we live, die, and that's the end—disturbing. It's not that we don't want to accept it, it's that it doesn't feel right. Something inside our bodies screams, "That's not right."

I'm certain that Lynda would state quite comfortably, albeit discreetly, that she was an atheist if indeed she was. In essence, Lynda encapsulates the humane face of faith—what the Bible is *really* all about, a humble and very personal feeling. She revealed to me that she has experienced the personal difference that faith makes in her life, and has observed the dramatic changes in the lives of individuals who practice their faith. Faith, Lynda feels, gives more than hope and belief—it gives direction, something lacking in the lives of many. I thank Lynda for allowing me to disclose her personal viewpoint, all for a good cause.

Lynda is typical of most scientists who believe in God. They do not thrust their beliefs upon others—if anything they will conceal them, knowing that they will be considered irrational. But why must logical people feel ashamed of having a strong sense of the existence of God? I hope that we can soon demonstrate that science and God are not mutually exclusive, but go hand in hand.

To make genuine ground in the science versus religion debate, you must look your adversaries in the face and consider that they may have contemplated this subject for decades. You should not consider your opinion to be superior to theirs from the outset. But you must also determine whether they are hiding a preconception. If they are, then their opinion and years of contemplation become worthless—they will only have developed their bias further.

The question I asked in this book, following some chance discoveries of new parallels between science and scripture, is: Does science prove the book of Genesis to be true? Chapters 2 to 8 have placed the opening page of Genesis under the microscope—phrase by phrase, creation event by creation event—and sought the alternative explanations for each event from science. It was perhaps unsurprising that the first event or two in the Genesis creation account matched the science perfectly. But as we progressed further through the verses, the parallels kept on coming—in every case there was a close match between the sequence of creation events in Genesis and the modern scientific account of how the universe and life came to be. This took me by surprise, to say the least. These later events on the first page of Genesis had even appeared rather abstract and meaningless as written, but when compared with the science, they suddenly made good sense. Here, then, is the Genesis Enigma: *The opening page of Genesis is scientifically accurate but was written long before the science was known. How did the writer of this page come to write this creation account?* From where did his thoughts, evidence, imagery, or inspiration come? The appendix that follows sets out what we know about this person, such as whether he was academically inclined. But I must admit, rather nervously as a scientist averse to entertaining such an idea, that the evidence that the writer of the opening page of the Bible was divinely inspired is strong. I have never before encountered such powerful, impartial evidence to suggest that the Bible is the product of divine inspiration. The Genesis Enigma may provide us with support for this proposition on a whole new level.

One's reaction to the science versus religion debate is a very personal choice. Do you believe that science will take such huge steps, changing the way in which it works today, as to be able to answer those big questions in the universe? Or do you choose God? To borrow from C. S. Lewis, do you believe that the whole universe is a mere mechanical dance of atoms, or that there is a great mysterious Force rolling on through the centuries and carrying you on its crest? Creationism and atheism are neither scientific theories nor demonstrably true. If we do not allow them to cloud our judgment then God can appear as a rational answer as to why we exist on earth.

APPENDIX: WHO WROTE GENESIS?

And so the Genesis Enigma has been revealed. Bizarre as it may seem, the stages of the creation account of Genesis 1 appear to measure up well against the stages in the scientific account of the origin of the universe and the diversity of life on earth. But how remarkable are the comparisons between each pair of equivalent stages in the ancient Israelite and modern scientific accounts? The answer to this question reveals the magnitude of the Genesis Enigma.

Was the creation story of Genesis 1 a consequence of its writer's cosmological and biological knowledge, in the way that we can consider today's account of evolutionary history as a product of our scientific understanding? Considering the depth and extent of scientific endeavor over the past few centuries that has been necessary to provide us with this understanding, we have assumed the answer to that question is "No." And if the writer of the Bible's opening page was without an understanding of the natural world, then the Genesis Enigma becomes almost unsolvable without the aid of divine intervention, in that it would be near impossible simply to guess the results of centuries of science. Yet to answer this question categorically, we need to know who that biblical writer was, and as much about him or her as possible.

Accordingly, this appendix will investigate who first penned the manuscript that turned into the heavily printed English blackface text on the first page of the 1613 Bible that I had aired in that fine London bookshop. This is a contentious subject within religious circles, but again, through discarding the primitive, literal interpretation of the Bible, we are provided with strong evidence to indicate who this writer really was. Using this unprejudiced approach, the writer of Genesis 1 can be revealed, although only through exploring the first *five* books of the Bible—the Pentateuch. But we must begin by dispelling the myth of Moses, the man traditionally acknowledged as the Pentateuch's author.

MOSES' PENTATEUCH

One day, somewhere between 1500 and 1200 BCE, in the Sinai Peninsula, Moses led the flocks of his father-in-law to new grazing ground, beginning a trend that the Bedouin of the area continue today. He found himself at the "Mountain of God," the present-day Mount Horeb or Sinai, where tradition claimed that divine beings resided, presumably due to mysterious phenomena that may have included volcanic activity. Without warning, a thorn bush ignited. Amid the flames appeared an angel, or other divine image of sorts, through which God spoke. Moses was instructed to move closer to the bush and to remove his sandals, following which God introduced himself. God had chosen Moses to lead the Israelites out of Egypt. Finally the fire extinguished itself and miraculously the bush was left without charring. The wood itself had not been burning.

God would speak to Moses on two other occasions—once from the mountain itself and once at the Tent of the Meeting. In the final year of his life Moses wrote up his life's adventures, and experiences with the divine, as best he could with just mortal language at his disposal. He chose to do this by placing his tale within the context of the history of the ancient Near East, because this way the people would assimilate his messages. Moses' manuscript is known as the Pentateuch, which as I have mentioned became the first five books of the Bible—Genesis, Exodus, Leviticus, Numbers, and Deuteronomy. Well, that's the traditional story, anyway.

As the archaeological evidence given in chapter 1 suggests, humans had sensed the existence of higher beings, or creators, for some millennia prior to Moses. But their attempts to fashion a religion out of such feelings were clumsy, built on the foundations of human imagination. Pagan religion had dominated the ancient Near East, where the forces of the universe were worshipped through anthropomorphic gods and goddesses. The God who revealed himself to Moses, however, announced himself as the one true God. Moses became the first conveyor of God's word.

The way that the Pentateuch is written leads one to sense the hand of Moses, although this is not spelled out. Indeed, many have defended—and continue to defend most vigorously—the Mosaic authorship. For the people who believe in this, the Pentateuch's status as divine and holy is endorsed through God's instruction to place it in the Ark of the Covenant—the sacred coffer built, according to the Bible, at the command of God. Viewing the Pentateuch in this way, it is clear that no sacred word should be doubted. But there's another way to interpret the Bible. Rather than reading literal truth from the biblical prose, general messages can be taken from the various fables—messages that relate to modern society and for which the precise wording of the text is irrelevant so long as the message gets through. God does not disappear if we take this alternative approach.

Science, of all subjects, can help here. It can point us toward this "symbolic" alternative. First of all, if it were to emerge that Moses was not the real writer of the Pentateuch, then this would add to the growing body of evidence that the Bible cannot be literally true, word for word. There must have been some degree of artistic license. In this appendix I will reveal strong evidence suggesting that Moses indeed did not write the Pentateuch, which will leave us free to get on with the task of finding the real author.

To get a sense of this detective story, we must try to imagine ourselves among the villagers of the Bible land three thousand years ago. While caught up in a city rush hour today, competing for a seat on the train home, it's difficult to contemplate a prophet announcing a call from God. Pavements should become dusty tracks; suits, robes; skyscrapers, stone houses; briefcases, goats. I hope that the following facts and evidence concerning the land and lives of the people in biblical times will help.

The land that matters to the origins of Genesis is that on the eastern coast of the Mediterranean Sea, stretching some hundred miles inland and two hundred miles north to south. Today Lebanon, Israel, Syria, and Jordan occupy this region. To the north is the picturesque Sea of Galilee, a vast freshwater lake, while to the south is the salt-saturated Dead Sea. These water bodies are connected by the Jordan River. The climate ranges from temperate to hot.

Over the past 150 years archaeological excavations in the Bible land have revealed the lifestyles of previous millennia. The floor plans of stone houses and indeed whole cities have reappeared, complete with underground tunnels and the cisterns of their advanced water systems. Many clay pots, such as those deposited in the Palestine Exploration Fund by Woolley and Lawrence, have been recovered, sometimes with traces of their original contents—wheat or animal bones. Objects of silver, gold, bronze, and iron have been found, including coins, swords, and scythes. Partially fossilized remains of the original flora and fauna can also be found. And then there are the ancient texts.

The language of the area was Hebrew, or one of the many other languages similar to Hebrew. Cuneiform—the first form of writing—and hieroglyphics may also be found in the region. Genesis was written in Hebrew, and at that time there was no Hebrew word for "religion." Religion was integrated into life in general, not split off into a separate compartment as it is today. It was part of planting seeds, farming crops, and then eating them. Making clay pots and washing clothes were either a part of God's will or an act of respect for him.

People wrote on papyrus, mainly, but also on clay tablets, leather, and plaster; they also carved words into stone (and even into wood, leather, bronze, gold, and silver). Today the ancient papyrus has mainly turned to dust; so the clay tablets in particular, again including those unearthed by Woolley and Lawrence, have proved the most enlightening.

Professional writers could be military officers or they could be civil scribes, associated with the elders or the judges. They could also be sacred scribes, and most of the remainder of this chapter will be devoted to four of these in particular. But before considering those four scribes, and with the social scene set, we need to know more about the history of the people of the Bible land.

HISTORY OF THE ANCIENT NEAR EAST

For the first centuries of biblical scholarship, the time between around 1500 and 1200 BCE was considered as the very beginnings of civilization. To these early scholars, the seed of culture was sown out of the darkness of an intellectual void. Modern archaeological studies in Mesopotamia (modern-day Iraq) suggest that, on the contrary, humans were organized and cultured to some extent as early as 8000 BCE, producing pots and anthropomorphic religious statues out of clay. So between then and 1200 BCE nearly seven thousand years of human cultural development had taken place. To put this in perspective, less than three thousand years have passed from the time Genesis was written to today. Out goes the theory that Genesis was written when it was because that's when the first cultured humans appeared on earth—those people who could be trusted with, or wanted to know, God's secret, or who were the first capable of constructing a sophisticated foundation faith. So the question remains, "Why was Genesis 1 written when it was, apparently in 1200 BCE?" The answer to this must surely provide clues to interpreting the creation account. But before moving on to the questions of Genesis' authorship, we must consider the human history within the ancient Near East from that time onward. It's important to digest this history because, as we will find out later, these times precede and influence the writer of Genesis 1—he came along a little later.

The story begins when Moses led the Israelites out of Egypt, where they had been slaves, into ancient Israel in around 1200 BCE. Various events, recorded in the Bible, occurred along the way, including Moses' encounters with God.

On this journey, Moses had been joined by his spokesman Aaron. The relationship of Moses to Aaron is unclear, but both men take on particular significance in the writing of the Pentateuch. The traditional view is that Aaron was Moses' elder brother. At least it is generally agreed that Aaron was the first high priest of Israel.

The Israelites divided up their new homeland into twelve territories, each occupied by its own Israelite tribe with a tribal leader. A thirteenth tribe existed also—the tribe of Levi—whose members were distributed among the cities of each of the twelve territories.

The Levites were a priestly group, though not all of them were priests.

Within each tribe, or within a group of tribes, there were individuals who stood out because of their personal qualities or position in society. These individuals became judges or priests. Judges could be male or female. They were also military leaders: The ancient Israelites lived through times of both war and peace. Priests were usually Levites and always male—this is all-important to discovering who wrote the Pentateuch. Priests served at religious sites and their office was hereditary. Their most significant role was in performing sacrifices, in return for which they received a portion of the sacrificial animal or other produce.

Another person of note within the Israelite community was the prophet. A prophet was one who had been called by God to deliver a message to the people. Since this could happen to anyone, "prophet" was not a trade—prophets could also be judges or priests. They could be male or female. The most famous prophet was Ezekiel, a priest.

The dominant religion in the Near East as a whole at this time was a pagan one in which people worshipped the natural elements—the sky, the sun, the sea, the storm wind—fertility and death. Animals were sacrificed by priests. Statues were made to remind the people of their gods. The chief god—El, a male—presided over the others in the manner of chairman. Each of the other gods and goddesses was linked to an element of nature.

In Israel there was just a single god—Yahweh, also a male. Yahweh was a ruler too, and not associated with an element of nature. The form of Yahweh appears to have escaped contemplation; only his actions—and his anger and mercy—are spoken of.

As well as the ebb and flow of war and peace, there were times when the political system was strong, and times when foreign influences dominated. The economic system fluctuated similarly—there were good times and bad.

Eventually a change took place in the political, communal, and military structure of the Israelites. At the time this came about Samuel—a judge, priest, and prophet—held perhaps the position of most authority, but his influence was restricted to the northern region of Israel that surrounded his city of Shiloh. Shiloh was a major religious center because it was home to the tabernacle (a sacred structure in the form of a tent made by spreading a fabric over a rectangular, wooden frame) that housed the Ark of the Covenant, containing the two stone tablets bearing the Ten Commandments delivered to Moses by God. Also, the priestly family at Shiloh were descendants of Moses. Then Shiloh and the northern area came under attack from the Philistines. The Israelites would need greater unity to counter such a force, and so Samuel's response was to crown the first king of Israel, King Saul. The period of the judges was over, replaced by the period of the monarchy.

King Saul united all the tribes of Israel, although the tribal leaders kept his power in check. Still, he found himself with an army composed of the combined troops of each tribe, which withstood the Philistine advance. But since religion was not separate from the military—or from anything else for that matter—to retain his authority he needed the support also of the chief priests and, predominantly, the prophets. Unfortunately Saul overstepped the mark between king and priest-prophet, and Samuel gave his support to a potential new king, David.

David was a hero from the tribe of Judah—the largest tribe—in the south of the Israelite land. He had married one of Saul's daughters, but soon became Saul's adversary. As David's popularity rose, he attracted the priests of Shiloh to his cause. Saul's response was to have all of these priests killed, although one managed to escape.

Eventually Saul died in battle with the Philistines, and David, along with Saul's son Ishbaal, resumed control of the Israelite land. Ishbaal was assassinated and David became king.

Much is written about David in the Old Testament. Not only did he lead an extraordinary life himself, and lay down enduring political foundations, he also left a long line of descendants on the Israelite throne. He began the Davidic dynasty.

David was careful not to show a bias toward the tribe of Judah, and this enabled him to unite the whole country. He captured the city of Jerusalem, hitherto occupied by the Jebusites and so not associated with any one of the Israelite tribes, and declared it his capital. He appointed *two* chief priests, one from the north and one from the south. The northern priest was Abiathar from the family of Moses—he who had escaped the wrath of Saul. The southern priest was Zadok, a descendent of Aaron, the first high priest of Israel. That members of the families of both Moses and Aaron held equally important, powerful positions was of exceptional importance to the construction of the Pentateuch—probably, indeed, the chief reason why it appears the way it does today.

David formed a national army. He built a globally significant empire that included the surrounding territories of Edom, Moab, Ammon, Syria, and maybe Phoenicia, with Jerusalem at its center. The most sacred object—the Ark of the Covenant—was brought to Jerusalem, and the two high priests relocated there.

David had many wives and many children. When he grew old, the succession to his throne became an issue. The chief contenders were Adonijah, David's eldest son, and Solomon, the son of his favorite wife, Bathsheba. The people of power were divided in their support, but of most importance was the rift between the two high priests. Abiathar, the descendant of Moses, backed Adonijah. Zadok, the descendant of Aaron, backed Solomon. Another Aaron-Moses divide. Solomon, with the support of the "Aaron" faction, won, and after David's death he ordered the execution of Adonijah and of Adonijah's general, Joab. Abiathar survived, but was expelled from Jerusalem. That city was now a stronghold for the family and followers of Aaron. Importantly, the Israelites involved in writing the Pentateuch lived from the time of David or Solomon.

King Solomon built upon the work of David, taking advantage of his empire's position at the gateway on the trade routes between Africa and Asia. His kingdom continued to prosper and even become rich. Among other buildings, including palaces, Solomon built a historically significant Temple in Jerusalem to house the Ark of the Covenant. The Temple was of modest size and was out of bounds to anyone except the priests, and so ceremonies and sacrifices took place at the entrance to the Temple. Inside the Temple were two rooms—an outer room called the Holy and a cubic inner sanctum called the Holy of Holies. The Holy of Holies—the room whose dimensions were replicated in the Sistine Chapel—contained two statues of sphinxes called cherubs, on which Yahweh was invisibly enthroned. Beneath the Cherubs' wings lay the ark.

When Solomon died and his less politically adept son Rehoboam became king, the north broke away from the south. The single country of David became two—Israel in the north, and Judah in the south. Rehoboam ruled Judah, and a man named Jeroboam—renowned as a high achiever from an early age—was chosen as king of Israel. The difference in religious life between Israel and Judah will help us to identify the writers of the Pentateuch.

The country of Judah held an advantage over Israel: It contained Jerusalem, the City of David, along with the Temple, the Ark of the Covenant, and the home of the chief priest for the religion of both kingdoms. The people of Israel had no choice but to make regular visits to Jerusalem in Judah—to the City of *David* and the Temple of *Solomon*. David and Solomon had favored the people of the south in Judah over the northerners of Israel. For the northern Israelites, something had to give.

Their king, Jeroboam, established a new version of the common religion, while continuing to worship their single god, known as Yahweh. Israel saw new religious centers, new holidays, new priests, and new symbols of the religion. Now the northern Israelites no longer had to cross the border of Judah to fulfill their religious observances.

Two cities replaced Jerusalem in Israel—Dan in the north of the country and Beth-El in the south. Where the religious symbol in Judah was a pair of golden cherubs, in Israel it became a pair of golden calves. The calf, or "young bull" in Hebrew, was a symbol of strength. One golden calf was placed in Dan, the other in Beth-El. Like the cherubs, they were not statues of gods but only the pedestals of the invisible god Yahweh.

The Near East continued to fluctuate between war and peace, and the territory of both Israel and Judah expanded into that of their neighbors and retracted again. But they were always relatively small kingdoms, and as such were vulnerable to the mighty Egypt and Assyria. Worse still for Israel, their monarchy was unstable. In fact it lasted only two hundred years until, in 722 BCE, Israel fell to Assyria. Its population was dispersed. Some fled to Judah while others were deported into the various sections of the Assyrian empire, to become known as the ten lost tribes of Israel. Judah, meanwhile, existed for another hundred years.

Here I will close the history book for a moment, because during the two hundred years that Israel and Judah existed side by side, there lived two writers of the Pentateuch. Yes, *two* of the writers, neither of whom was Moses.

What is important to us for the purposes of this book is the world of the *real* writer of the Genesis creation account—where and when he or she lived, his or her job and lifestyle, his or her religious and political beliefs. With the myth of the Mosaic authorship dispelled, we can begin to make use of that human history of the Near East. Using the text of the Pentateuch itself, we can deduce what type of person would want to write each passage. Then we can try to find them somewhere in the Bible itself.

WHO WROTE THE PENTATEUCH?

Ultimately we would like to know the writer of the creation account of Genesis 1, but for this inquiry to become realistic, we must first consider the entire Pentateuch. The text of the Pentateuch—the first five books of the Bible—can be divided into four separate texts, written predominantly by four different writers. So why not simply consider only the writer of the first book, Genesis? Because the complete book of Genesis did not have a sole writer. The four texts do not appear as continuous, consecutive stories. No, the four texts were literally cut up and pasted together by an editor, the "redactor." The reasons why this was done, and the explanation of how we know this, are wrapped up in the history of Israel, as I began to summarize above and will continue to do below. So, to identify the writer of Genesis chapter 1, we must first identify his or her work as it appears spread throughout the entire Pentateuch, which means at the same time distinguishing the work of the other writers, too.

For centuries, biblical scholars had recognized different styles of writing in the Pentateuch. At first they were persecuted for meddling with the divine words of Moses. But when the same conclusions were reached within different monasteries, the idea that Moses was in fact not the writer began to take hold. Clearly, some paragraphs scattered throughout the Pentateuch shared the same style of prose, and the same political biases and opinions, while others differed markedly in both respects.

Just as the author of a murder mystery would involve the most sophisticated forensic tools of their time, to gain repute and popularity, and so that readers could relate to them, the writers of Genesis placed their messages within the context of their own times. Since they lived in different centuries, and had different points of view, their prose, styles, and allegiances were different and can be distinguished from one another today. Often, more than one writer told the same story, and so the various versions were chopped up and reassembled as a single composite by a person known as the "redactor." This may seem counterintuitive, but this method served its purpose in that in each case it produces a version of the story that represents the viewpoints of different individuals and generations. It also evened out the prejudices.

The idea of multiple authorship of the Pentateuch first came to light when in 1711 a German minister, Henning Bernhard Winter, noticed that the same stories recurred, and that each version had its own unique characteristics. For instance, the story of the flood and Noah's Ark appears twice, but in one version the creator is referred to exclusively as "God," while in the other he is given only his name, "Yahweh." Someone, Winter concluded, must have fused two separate accounts of the same flood story into a single document. His work was overlooked, as was that of the French professor Jean Astruc, who reached the same conclusion in 1753. But when a third person exposed this fact—the respected German scholar Johann Gottfried Eichhorn, in 1780—the world was obliged to take note. Eichhorn named the group of stories that referred to "God" in the Pentateuch as "E," because the Hebrew word for God is "El" or "Elohim," and the stories using "Yahweh" as "J," which in German is pronounced like the English "Y." Eighteen years later it was discovered, on the basis of differences in vocabulary, that these "two" writers of the Pentateuch were in fact four. Still, some today prefer to cling to the old tradition that Moses wrote the entire Pentateuch.

Taking the progressive or "documentary" approach, the four sources of writing evident in the Pentateuch were given the labels J, E, P, and D, where "P" stood for the priestly source and "D" for Deuteronomy. In addition, a single editor, known as the redactor, merged the sources. The writers did not know each other or the redactor, and the four texts were composed during different periods in time. As a result, the Pentateuch emerged as one of the most extraordinary texts ever composed, believed by many to be greater than the sum of its parts.

From the condensed history of the Israelites presented so far, we can identify the writers of the J and E sources. Throughout this appendix I will refer to the modern biblical scholarship of Richard Elliott Friedman, a professor of Hebrew at the University of California, San Diego. In my less than authoritative opinion, Professor Friedman's work provides a judicious integration of the many theories that crowd the subject of biblical authorship. So let's look at some of the evidence that reveals the writers' identities.

SOURCES J AND E

I mentioned that two of the writers of the Pentateuch lived between 922 and 722 BCE. These were the writers of the J and E sources. The writer of J lived in Judah, and the writer of E in Israel. Each composed a version of the same divine message uniquely dressed in his own people's clothes. Not surprisingly, the two stories differ noticeably.

As we begin to single out the differences between stories, the personalities and loyalties of the writers begin to emerge. One of the sources of the flood story was J, while the other was P (P will be considered in detail later). In addition to using the name "Yahweh," J includes not a pair of each kind of animal as P does, but seven pairs of clean animals and one pair of unclean animals. "Clean" means fit for sacrifice. Sheep are clean, elephants are unclean. From this, we can assume that the writer of J had more than a passing interest in sacrifices.

The J source further states that the flood lasted forty days and forty nights, but omits any details of ages, dates, and measurements in cubits. Its writer has no head for figures, except for the number forty, the number relevant to Moses in the desert. Most typically, the deity in J is one who can regret things he has done. Further, he has a heart and a nose—he has human qualities. The writer of J is more the poet than the engineer. These characteristics are consistent with all the J stories in the Pentateuch, which include the end of the Creation account (after Adam and Eve), the Garden of Eden, Cain and Abel, the Generations of Man, Sons of God, and human women . . . the list is long.

The E source, on the other hand, calls the true God "El" or "Elohim" until after the arrival of Moses. Then, in a scene in Exodus, God reveals his name as Yahweh, and the E source refers to "Yahweh" thereafter. The E source is the last to emerge in the Pentateuch, first appearing in chapters 20 and 21 of Genesis with the birth of Isaac. A little further on in the Pentateuch, E provides an account of Jacob, and from here onward the J and E stories overlap considerably.

If we place the J and E versions of the same stories side by side, patterns begin to emerge. Alongside consistencies within each source, differences in opinions between the writers of the two sources soon surface to reflect two different ways of life. The first difference is the geographical setting of the stories. The J stories place the home of the patriarch Abraham in Hebron. Hebron was the capital of Judah under King David. The J stories continue to justify the ascendancy of Judah. Then, in one J story, Yahweh promises that Abraham's descendants will have the land "from the river of Egypt to the . . . river Euphrates." This is the nation's land under David. Clearly the writer of the J stories lived in Judah rather than Israel. Yet in the E stories, Abraham's grandson Jacob had a fight with someone who turns out to be God or an angel, and this fight took place in Peni-El, a city in Israel. In fact the E source has a plethora of stories suggesting that its writer was from Israel. A boundary conflict is clear where both writers set stories in Beth-El, a city on which both Judah and Israel made political claims.

Another noticeable difference between the J and E sources occurs in the way they refer to the various tribes of Israel. The E source includes the names of all the tribes of Israel. The J source includes only the three tribes who lost their territory and merged into other tribes, including the Levites. In addition, the J source specifically names the tribe of Judah—the only tribe with existing territory cited.

The J and E sources contain clues not only to the nationality of their writers and the time in which they lived, but also to where their interests lay; and from here we can begin to deduce their professions and, ultimately, their identities. The story of the golden calf is particularly revealing of the E source's writer, so I will use this and one or two others to provide a flavor of how we can extract clues.

While Moses was on Mount Sinai receiving the Ten Commandments, Aaron made a golden calf for the people, who replied, "These are your gods, Israel, that brought you up from the land of Egypt." Aaron declared a holiday to Yahweh "tomorrow," and the people celebrated and sacrificed avidly. God told Moses what was happening, and said that He would destroy the people and start a new community descended from Moses. Moses diffused the tension but, with his assistant Joshua at his side, in his anger he smashed the tablets bearing the Commandments. The tribe of Levi gathered around Moses and carried out a bloody cleansing among the people. Moses pleaded with God to forgive the people and not to destroy them. This story holds clues to the social and "political" positions of the writer of the E source. Moses is cast in a positive light, while Aaron is the heretic, although he receives no punishment. It is clear that the writer of the E source was from Israel, and owed allegiance to Moses.

Curiously, Joshua is included in the story, although distanced from the heretical golden calf event. In other E stories, Joshua is also Moses' faithful assistant, who serves in battle and prevents the misuse of prophecy. Joshua plays no role in the J source. This is because Joshua was, as we might by now expect, a northern hero, coming from the tribe of Ephraim—Jeroboam's tribe.

The story contains yet more clues to its writer's identity. Further on in the golden calf story, the Levites (apart from Aaron) are pictured as acting in bloody fanaticism. They were acting against Aaron, and were the only group among the Israelites to act out of loyalty to Moses. This suggests the identity of the writer as a Levite himself. Then Aaron is shown to contradict himself by declaring a holiday to Yahweh while offering the golden calf as an alternative god—in other words, he had "mistakenly" invoked *two* gods within the Israelites' monotheistic religion. Nevertheless, criticisms of Aaron himself are disguised, not overt.

Strangely, the E source describes the single golden calf as "gods," in the plural. This was simply to gain authority. In the account of King Jeroboam in the First Book of Kings, Jeroboam made two golden calves and declared to his people, "Here are your gods, Israel, that brought you up from the land of Egypt." In the golden calf story in the E source, the people also proclaimed that the golden calf "brought them up from the land of Egypt." Clearly the writer of the E source was invoking words that had been traditionally ascribed to Jeroboam, to make a strong connection between his golden calf story and the golden calves of the kingdom of Israel. Just as clearly, then, Jeroboam was the king who mattered to the writer of this story.

Remember, in ancient Israelite history, Jeroboam had established an independent kingdom in the north—Israel, as opposed to Judah. Well, we have seen already that the writer of the E source came from Israel. But his golden calf story is not exactly a happy one. For instance, the Israelites are depicted as rebellious immediately after their liberation from Egypt. So this Levite priest from Israel did not have a good time under Jeroboam. Their time of liberation had been turned into a time of religious betrayal. Referring again to the history of ancient Israel, one group of Levites were known to have suffered at this time—the Levites of Shiloh. These were exclusively male, and were supposedly descendants of Moses.

In summary, we can deduce that the writer of the Pentateuch's E source was a man, a Levite priest of the city of Shiloh, in the northern kingdom of Israel, from the family of Moses. This is also the conclusion drawn from many other E stories.

In a J story, Moses sent a group of spies from the wilderness into the "promised land." The spies had traveled through the Negev, the scene of Woolley and Lawrence's early expedition described in chapter 1. They traveled as far as Hebron, at the southern end of their territory, then to Wadi Eshkol, still within Judah's borders. All except one of the spies warned that this was a land of fierce warriors. The one exception was the spy Caleb, who actually encouraged the people to have faith in the land. The people listened, and Caleb became the founder of the Calebites, those who held territory in the hill country that included Hebron, Judah's capital. The important point about this story is that, like many other J stories, the setting was Judah. The spies reached its further borders, but never crossed them. Judah was the J source's promised land.

So the E and J sources cover a similar period of ancient Israelite history, in many cases overlapping in their stories and characters. That's either because they tell, independently, the traditional stories of the Israelite forefathers—a common treasury of history and tradition—possibly using an earlier written version, or because one was based on the other. The writers certainly carry opposite allegiances, which is because E heralded from Israel, and J from Judah. If the E texts were produced first in Israel, then Judah would certainly require an equivalent text with the emphasis shifted toward different heroes, and vice versa. Indeed, if we are told where they derive from, and so where the writers' loyalties lay, it becomes obvious why the E and J sources were written in the way they were.

For instance, the J source mentions the Ark of the Covenant at every opportunity, while it is never mentioned in E. That's because the Ark of the Covenant was held in the Temple of Solomon in Jerusalem, Judah. Conversely, the E source waxes lyrical about the Tent of the Meeting, or Tabernacle, associated originally with the city of Shiloh in Israel. The Tent of the Meeting is absent from the J source.

The J source reflects the interests of the Judean court, where both men and women had a certain status. Although society was male-led, noble women had more power and education than males of a lower class. In general, the J stories are much more concerned with women than the E stories. For instance, the Tamar story in Genesis is sympathetic to a wrong done by a man to a woman named Tamar. At the end of this story, the man acknowledges Tamar's rights and his own fault. This opens the possibility that the writer of the J source was female.

We can be sure about the time in which the J and E sources were written. As I mentioned earlier, they were certainly both written before 722 BCE, when the Assyrians destroyed Israel and exiled its population. It is clear from the stories that this had not happened when the J and E sources were written. We can also date the events that were written about. In the J source the Ark of the Covenant has central importance, and so the writer must have lived after the division of the kingdoms of Israel. Then, the J stories of Jacob and Esau reflect Edom's independence from Judah, which occurred during the reign of the Judean king Jehoram, between 848 and 722 BCE. So the writer of J must have lived in this period. Unfortunately it is difficult to narrow down the interval in which the E source was composed, so we are left with the period when Israel stood, between 922 and 722 BCE. There is still more work to be done toward pinpointing the dates of the E and J sources, and in delving further into their precise lifestyles. But what we do know of them so far is still extremely useful to making sense of the Pentateuch.

Remember that Israel fell in 722 BCE, but not Judah. It is likely that the E text came to Judah with the flow of people from Israel at around this time. Inevitably, the E and J texts would have come together, to fashion the "JE" document. Both texts were sufficiently well known among the ancient Israelites that, like the people themselves, an amalgamation would have been unavoidable.

We know how the history of the Bible land continued after the fall of Israel. A political change took place that also meant economic and social change and, as always, religious change. This affected the development of the Pentateuch, and so we need to know something about it.

MORE HISTORY OF THE ANCIENT NEAR EAST

After 722 BCE the Israelites were all crowded into a territory half the size of that ruled by David and Solomon: Judah, which now operated from a position of weakness. This was the age of Judah's powerful neighbors: first Assyria, then Babylonia. Both had ambitions to engulf Judah, and this fact was central to Judah's politics.

Statues of Assyrian gods were placed within Judah's temples as a necessary sign of goodwill, although some religious devotees might have seen this as heresy. Indeed, some devotees fought Assyrians in Jerusalem when Assyria tried to exert an influence. On the minds of *all* Israelites, meanwhile, was the awareness that Israel had fallen while Judah lived on. This was interpreted either as God's preference for Judah, or as a warning to Judah that any kingdom could fall. The separate tribes of Israel had all but died out, but the priestly groups remained in existence and could have come into conflict on the meaning of Israel's demise. Surely, any northern Levite priests arriving in Judah after 722 BCE would have brought opposing views to the table. We must consider this scenario as we pursue the remaining two writers of the Pentateuch—those behind the D and P sources. We must also consider a change to religion. For by now, for the first time, a religious book existed—the JE document. This was to influence the completion of the Pentateuch.

King Hezekiah, who ruled Judah from around 715 to 687 BCE, introduced both religious and political reforms. Religious practice became centralized at the Temple in Jerusalem, with the various other forms of worship declared illegitimate. Political practice became assertive, with insurgence against Assyria, and an attempt to extend into the new Assyrian territory in the former Israel.

The Assyrians hit back, bombarding Judah and capturing strategic cities. But although crippled, Judah did not fold as Israel had done. Decisively, the key city of Jerusalem remained an Israelite—or at this stage more appropriately a "Judean"—stronghold. Somehow, the Judeans withstood the Assyrian siege of Jerusalem, aided no doubt by the hilltop location of the city, but also by Hezekiah's tunnel (the Siloam tunnel) underneath it through which vital water supplies could be drawn from the spring below. This tunnel, an architectural wonder, remains in existence today, albeit only as a tourist attraction.

The population of Jerusalem grew. The Judeans of the countryside were compelled to cross the land to bring their sacrifices to Jerusalem. There would have been a great flow of livestock and produce to the city under Hezekiah. When Hezekiah died in around 687 BCE, the people buried him in the best part of the necropolis that held the sepulchres of the kings, in tribute to his great honor and devotion to God.

Hezekiah's son Manasseh, and then his grandson Amon, followed as rulers in Jerusalem, although in a different manner. They were forced to reintroduce pagan worship in Judah, probably as a result of Assyrian pressure. But King Amon was killed when he was twenty-four, and his son Josiah succeeded him at the age of just eight. At the beginning of King Josiah's reign, government would have been in the hands of priests serving as regents. Then, when he became old enough to rule, Josiah reintroduced Hezekiah's religious policies. Pagan idols were smashed, and the Temple in Jerusalem was cleansed and proclaimed the center of religion. The priests of Judah were once more obliged to traipse the land and return to Jerusalem, even if it meant accepting second-level jobs beside the Temple priests.

Josiah had already felt the various influences of his court, the priestly circle, domestic politics, and Assyrian pressure when in 622 BCE, in the eighteenth year of his reign, a book arrived on the scene.

Josiah's scribe Shaphan reported that the priest Hilkiah had uncovered a "scroll of the torah" (torah here meaning "teaching"; not necessarily "the Torah," that is, the complete Pentateuch) in the Temple of Yahweh in Jerusalem. This book was read to Josiah, who reacted with extreme anguish, and held a giant national ceremony to affirm the renewal of the covenant between God and the people. Further, Josiah destroyed the altar at Beth-El where King Jeroboam's golden calves had once stood, and in doing so proclaimed his interest in the land that was once the kingdom of Israel. More about the book that triggered this conduct later.

As the years passed, Babylon was rising and Assyria weakening. Egypt sent forces through Judea to support Assyria against the Babylonians. Josiah confronted the army and, at the age of forty, was killed by an Egyptian arrow. As it turned out, the Egyptian aid was in vain, and the Babylonians overpowered the Assyrians. In the years that followed Judea was ruled by Josiah's sons, who were in turn followed by an Egyptian proxy by the name of Jehoiachin. Then Babylon reasserted its dominance, replacing Jehoiachin on the throne with another of Josiah's sons—Zedekiah.

Zedekiah ruled Judah for eleven years. In his ninth year of rule he rebelled against the Babylonian king Nebuchadnezzar. This was a bad idea. The Babylonian army returned and destroyed Jerusalem, and thousands of Judeans were exiled to Babylon. Nebuchadnezzar executed Zedekiah's sons in front of him, then immediately blinded Zedekiah himself. Thus the rule of King David's family in Jerusalem came to a horrific end.

Nebuchadnezzar placed on Judea's throne Gedaliah—a member of a pro-Babylonian family that included the famous prophet Jeremiah. When he was killed two months later by a member of the house of David, the Judeans who remained in Judah so feared Nebuchadnezzar's response that they took their only option—they fled to Egypt. This was a sad irony, taking them back to the land where they had been enslaved before their Exodus to the promised land (under Moses in the thirteenth century BCE).

In 587 BCE Nebuchadnezzar captured and burned Jerusalem. The city, including the Temple, was destroyed. The Ark of the Covenant was lost, and its fate remains a mystery today. Judah's 400-year-old royal family had been dethroned, and the Judeans were exiled either as captives in Babylon or as refugees in Egypt.

We can almost place the history book on hold for now, since we have covered the period of the D and P sources, with one factual exception. In 516 BCE another temple—the Second Temple—was constructed in Jerusalem to replace the Temple of Solomon that had been destroyed. This stood as the center of worship until it was burned to the ground by the Romans in 70 BCE. The writers of the D and P sources lived before the Second Temple period, but it is useful to know this. Now to consider how the writers living in the times above would have experienced their world. It was certainly a very different world from that of the individuals who wrote the J and E sources. Their sense of God would have differed too.

SOURCE D

The book that the priest Hilkiah said he found in the Temple in 622 BCE was the D source—a near-complete version of Deuteronomy, the last book of the Pentateuch.

Deuteronomy reads as Moses' farewell speech before his death. It is set in Moab, to the east of the Jordan river, where Moses and the people arrived after their forty years in the wilderness. Moses gives the people a code of law by which to live their lives in the future. He appoints Joshua as his successor, then climbs to the top of a mountain and dies.

The biggest clue to identifying the writer of the D source is that he also wrote the next seven books of the Bible to follow the Pentateuch—Joshua, Judges, Ruth, 1 and 2 Samuel, and 1 and 2 Kings. Deuteronomy as compiled at a later stage by the redactor is based chiefly on the D source, along with minor additions from other sources, including E and P. The writer of the D source, and the next six books of the Bible, used available older texts to help carefully compose his history of the people of Israel and their land. This includes the divisions of the land, and changes in monarchy and religious and administrative practices. Actually, the D source can itself be divided into two parts—the first, written before 587 BCE, covering about 95 percent of the book, and the second, which must have been written after this date. The history covered in the combined D source stretches from the time of Moses to just after the end of Judah—which came about, as we have just seen, in 587 BCE.

The D source places considerable emphasis on covenant. The fate of kings and the people appears dependent upon how faithfully they kept their covenant with God. The idea of covenant is of paramount importance to the Bible as a whole—indeed, the Latin word *testamentum*, as in the titles "Old Testament" and "New Testament," means "covenant." Hence the "Old Testament" is the description of the Jewish people's covenant with God before Jesus. The D source makes this message clear.

In the D source, "the covenant" includes the laws that Yahweh gives to Moses at the end of the forty years in the discipline—the laws of Deuteronomy. Yet later in the D source, another covenant is introduced. This is revealing of the writer's identity. It is a covenant between God and King David.

God promises David that, as a reward for his loyalty, his descendants will rule the kingdom forever. This provides an indication both of the writer's allegiance and of the time that he was writing.

He covered the history of the kingdom up to and including the rule of King Josiah, and predicted that the Davidic family would continue to rule forever. He could not have known about the end of that rule. He must, therefore, have written the first part of D before 587 BCE, and after Josiah came to the throne in 604 BCE. Further, we can deduce from the D text itself that he lived in Judah, the land of King David.

After 587 BCE, a story predicting the eternal rule of the Davidic family would have appeared a little odd, if not imperfect—an adjective not to be associated with religion. It would have been quickly cast aside as a sacred text. This is why the second, smaller part of the D source was written, soon after 587 BCE—to correct a prediction so glaringly inaccurate.

The D source deals extensively with Josiah, and Josiah's sons are not mentioned at all. This and other clues in Deuteronomy indicate that D was written during the rule of Josiah. But a little more on the reference to Josiah.

Josiah, along with Moses, is revered most highly in D. He is placed on a level with Moses, with the assertion that "none arose like him" made of both, but no others. Then, Josiah's existence is predicted by name during King Jeroboam's visit to the altar at Beth-El, three hundred years earlier. "Here a son will be born to the house of David, Josiah by name," says Yahweh. Even in a book packed with prophecies and miracles, there is no other case of such explicit prediction of a person by name so far in advance. It's as if the covenant that began with Moses would be fulfilled by Josiah, where the curtain falls on the religious tale: opened by Moses, closed by Josiah.

The law code occupies about half of the book of Deuteronomy. This includes the "Law of the Kings," which places restrictions on a king's activities. There is a surprising instruction that a king must write a copy of these laws and read it regularly in front of the Levites. In fact the Levites appear often in the law code, and they are always favored. So it is likely that the D source was written not by someone in the royal court, but rather by a Levite priest. But from which Levites?

The first of the Deuteronomic laws decrees the centralization of religious practice—that all sacrifices be brought to one central altar. This discriminates against the rural Levites, and indeed it was the law that put them out of business, although at the same time it is sympathetic toward them by further instructing the people to care for them. But only the interests of a group of central Levites are actually *promoted*.

Not surprisingly for a Levite text, the D source gives no attention to anything associated with Aaron (the brother of Moses). This includes most religious artifacts that were housed in the Jerusalem Temple. So the Levites in question favored centralization of the religion in Jerusalem, but were not followers of Aaron. The evidence points to only one group of Levites—the priests of Shiloh, the same group that produced E.

Another clue pointing to this Levite group is that their law codes did not favor the priests who served at Beth-El before the fall of Israel in 722 BCE. This is because the Levites of Shiloh had been excluded by non-Levites at Beth-El. The Levites of Shiloh refused to accept the central significance of the ark and the Jerusalem priesthood because their leader Abiathar had been expelled from Jerusalem by Solomon, and the Jerusalem priesthood had been Aaronid ever since. They accepted that there should be a king, because their leader Samuel had proclaimed the first two kings of Israel, but only one with limitations on his rule, because the kings Solomon and Jeroboam had treated them badly. They preferred tribal forces to a professional army because the rise of professional armies led to the independent power of the kings, no longer dependent on the support of the people.

Certainly, the law code of Deuteronomy reflected the interests of the Shiloh priests. It is interesting that the law code appears in the middle of this book. It would seem that its writer later penned an introduction in the form of Moses opening his last speech and reviewing his forty years with the people. Then he added to the law code a list of blessings for those who acted well, and curses for wrongdoers, and finally concluded D with Moses' last words and acts, including his exhortations to the people. Interestingly, Moses is described as having written this book—this "torah"—on a scroll and having given it to the Levites with instructions to place it next to the ark.

The writer of the D source continued his work on the history of Israel, following the death of Moses, through the Old Testament books of Joshua, Judges, Ruth, Samuel, and Kings. The grand finale sees the priest Hilkiah finding a scroll and King Josiah acting upon it.

Now think back to the history of ancient Israel, to 622 BCE, when Josiah received a book that caused him enormous distress. Josiah's scribe Shaphan had reported the discovery of this book by the priest Hilkiah in the Temple in Jerusalem. Well, that book and the D source we are discussing here were one and the same. One can imagine Josiah's emotions when those Deuteronomy laws were read to him, setting out the restrictions on his rule. He destroyed the altar at Beth-El where King Jeroboam's golden calves had once stood. That's because the first Deuteronomic law code demands a single place for sacrifice, at Jerusalem, and so Jeroboam's establishment of the golden calves at Beth-El and Dan suddenly appeared a dreadful sin. Josiah's further reactions to this book saw him taking more interest in the former kingdom of Israel, as a result of the positive light that D had cast on this land—the land of Moses. Incidentally, these actions, and indeed the D source itself, were all favorable to the Levite priest Hilkiah.

So how did this D source—this torah of Moses—find its way into the Temple in Judah, and into the hands of Hilkiah? To answer this, we must consider one other person connected to King Josiah and the D source—the prophet Jeremiah.

Jeremiah was a sincere and righteous man. He genuinely believed that what he wrote was true. He wished that he could not see the future, but that was, alas, his destiny, which he could not escape even by death. He was a solitary, tortured man, rejected by other humans, who feared and persecuted him. This picture is painted in the book of Jeremiah, which appears later in the Old Testament and also reveals that Jeremiah both lived alongside and admired King Josiah. Independently, the Old Testament book of 2 Chronicles indicates that Josiah composed a lamentation for Jeremiah when he was killed. More precisely, Jeremiah can be connected with Hilkiah, the priest who discovered the D source, and with Shaphan, the scribe who carried it to King Josiah and read it to him. A son of Hilkiah and two sons of Shaphan stood by Jeremiah during critical moments in his life, for example to save him from being stoned. Interestingly, Jeremiah's father was a priest named Hilkiah, although we cannot be sure that this was the same Hilkiah who found the D source.

Jeremiah was a Levite priest from Shiloh. We know this because, while frequent reference is made to his priestly status, he never sacrificed; this is consistent with the practice of the priests of Shiloh. Then, he referred to Shiloh as the great central place of old. Further, Jeremiah was the only prophet to mention the story of Moses' bronze snake, which derived from the E source from Shiloh.

That Jeremiah was an acquaintance of Hilkiah and Shaphan, the men who first handled the D source scroll, and belonged to precisely the same sect as the writer, and that the scroll was written during Jeremiah's lifetime—all this is perhaps too much of a coincidence. Another piece of evidence makes the connection even more robust:

The books of Jeremiah and Deuteronomy employ the same language and outlook.

Some parts of the books of Jeremiah and Deuteronomy are so similar that the same person must have written them. Grammatical features, special terminology, themes, syntax, and literary structure all match. Since Jeremiah wrote the book of Jeremiah, then he is the author we seek for D. Jeremiah wrote at least most of Deuteronomy—along with most of the next seven books of the Bible. He probably wrote it just before he handed it to Hilkiah in 622 BCE.

Before moving on to the final source we have to explore, it is worth considering how Deuteronomy had been shaped by the history of ancient Israel. After the fall of David's family, the emphasis in Deuteronomy shifts to the covenant that Moses mediated between God and the people. This is because any hopes the people might have for security could no longer be based on the Davidic covenant. The Davidic covenant became a promise only that the throne was eternally available to David's family. Even if it was unoccupied at the present, the possibility existed that one day a descendant of David—a messiah—might come and rule justly. Of course, the consequences of that possibility for Judaism, and for Christianity, were immense.

SOURCE P

It has been interesting to discover how the Pentateuch can be attributed to four writers, and how we can isolate a writer predominantly by using the biblical text itself. Well, so far we have identified three of those writers, either as a single individual or as a member of a particular sect. In the case of the E, J, and D sources we have seen how the allegiances and prejudices of their writers were reflected in the texts—so strongly that there was no doubt as to who must have written them. In reverse, once we know their identities, it becomes clear why they wrote what they did. This is good to know, because maybe we can apply the same logic to the final writer. It is also good to know that each author independently penned a history of Israel that is like all the others, except for details affected by opposing allegiances, and so that they were not in the business of constructing myths. All three of the authors considered so far faithfully recorded what they knew about their people's history, each under the influence of his or her particular cultural and natural environment. The redactor must have enforced a ruling that only such reliable texts could enter the sacred Pentateuch.

The last of the writers we need to identify, who strangely enough is the last to emerge during the scholarly search for the Pentateuch's writers, is the one we are most interested in—the writer of P. The P source must also have passed the redactor's test, and so we can assume that it follows its author's genuine impression of ancient Israelite history. Would it not be interesting if this writer's identity and outlook made his reason for writing the creation account in Genesis 1 self-evident? Traditionally, the search for the P writer has been the hardest of the four tasks. P is the dominant source in the Pentateuch—equal in size to E, J, and D combined. Though it includes a great deal of law, it is essentially a narrative, recording the physical history of the earth, and the saga of the descendants of Abraham, Isaac, and Jacob. As such, it contains many stories of apparently real people, and accounts of many of their adventures, including the Exodus and the Israelites' journey through the desert, that, like many other of the P stories, are alternative versions of stories in J and E. It begins, nonetheless, with the creation account that made it onto the Bible's opening page.

The P writer was strict on details, in the manner of an accountant today. He described the minutiae of the booty taken, and its distribution, during war, and of censuses during peacetime. But the P source is without dates—time is deduced from chronological statements and genealogies. Yet for all its austerity, its simple elegance has attracted literary praise. Its style borders at times on the poetic, and the relative proportions of its verses given to different themes and episodes are a clear sign that it had been well thought out. P was written by someone keen to record a story using the facts, with the literary skill and intelligence to make his text appealing to the masses.

We should begin the search for the writer of P by narrowing the period in which it was written, and by considering the texts that it used, and those that used it. These are texts for which we do have dates. To begin with the texts that it used, the P source clearly counters the J and E stories. Indeed, it appears an alternative to JE in many parts. It provides alternative accounts of the flood, Moses and the Ten Commandments, and of many more events including the creation. But it adds stories of subsequent events in history, as I will mention below, which clearly indicates that P was written after J and E rather than the other way around. The most conservative, latest date we have for the production of J or E is 848 BCE.

Now to consider texts that use the P source.

Ezekiel was a priest living with the Israelite exiles in Babylon in around 600 to 580 BCE, after the fall of Jerusalem. The book Ezekiel wrote in Babylon, which came to be included in the Old Testament, includes many stories in which words are taken from the mouths of P source characters and expanded upon. For instance, the story of a contract in the P source is continued to cover a lawsuit relating to that contract in the book of Ezekiel. The book of Ezekiel must have been written after the P source; and P had to have been written by 580 BCE. But there's another, earlier writer who must have had access to P—the writer of D, Jeremiah (or, more specifically, his scribe).

The D source refers to the Moses stories in JE—as we would expect, Jeremiah had access to the JE scroll. But there's an exception. One of the Moses stories in D is missing from JE, but can be found in P. This is the spy story (involving the spy Caleb discussed earlier), and Jeremiah quotes the P version word for word using the linguistic style of the rest of the P source. This means that Jeremiah also had P in his library. We know that D was written in around 622 BCE. It follows that P was written before this. This provides us with a new, conservative date for P of between about 848 BCE and 622 BCE. Now to scour the *contents* of P for further clues to narrow this window, and thereby to close in on the identity of the author.

First, the P source refers to the fall of Israel, and so must have been written after 722 BCE. It is equally striking that P must have been composed before the fall of Jerusalem in 587 BCE. It insists that sacrifices should take place centrally, only at the Tabernacle, or the Tent of the Meeting as it is otherwise known. These were the specific names for the tabernacle held in the "First" Temple—the Temple of Solomon, in Jerusalem. We know already that the Tabernacle was destroyed with the First Temple in 587 BCE. As I have mentioned, a Second Temple was constructed in 516 BCE, but this certainly would never have been associated with the Tabernacle or Tent of the Meeting. P's insistence on the use of the Tabernacle indicates a date between 722 BCE and about 587 BCE. But there's further information to be gleaned from the text of P.

The text of the P source is specifically loyal to one king—Hezekiah, Josiah's great-grandfather. It emphasizes the centralization of religion, including one place for sacrifice. This act originated with King Hezekiah. It also speaks of the priests and the Levites as distinct groups—again, a division brought about by Hezekiah. This and other evidence suggests that P was written during Hezekiah's reign. Scholarly opinions on the date of this reign are broadly in agreement. Two authorities on the subject, for example, William F. Albright and E. R. Thiele, proposed 715 to 687 BCE and 716 to 687 BCE, respectively.

Integrating all the periods deduced for P so far, we are left with a twenty-eight-year window within which P was written—between 715 BCE and 687 BCE. Although some of the dates calculated for P above have proved less useful, they do at least all support this period. So, with an approximate time of writing in place, we can turn our attention to the *identity* of the P writer.

The text of P is very specific about who was eligible for the priesthood, namely Levites who were descended from Aaron. It seems that the P writer was either an Aaronid priest himself, or someone representing the interests of Aaronid priests. His loyalty to King Hezekiah further attests to this. King Hezekiah smashed the bronze serpent that, according to E, was made by Moses to cure deadly snakebites. Moses was the patron of those Levites who were rivals to the Aaronid priests at the time. Clearly, Hezekiah favored the lineage of Aaron. This must have been a good time for the Aaronid priests—their best since the days of King Solomon.

As an Aaronid priest, the writer of P was most probably male. It is clear from his writings that he was very familiar with priestly practices in Jerusalem, and knew the Tabernacle and its literature in some detail. For instance, he describes the materials involved in making the tent itself—frames and poles of precious wood, gold and brass rings, and sheets of wool and linen fabric woven with gold, scarlet, and purple, with a covering of red leather. This is most probably because he lived in Jerusalem.

To summarize, *the writer of the creation account of Genesis 1 was a male, an Aaronid priest who worked in Jerusalem between 715 BCE and 687 BCE*. That we do not know his name is unimportant. The information that we *do* have could point to the P writer's reason for setting out his creation account in the way that he did. It should be noted that some scholars suggest the P author did insert some work of others into his document. But, if true, this relates only to the P *laws*, and does not affect the creation account of P whatsoever.

As an Aaronid priest from Jerusalem, the P writer must have been unimpressed by the parts of the JE book that derived from the E source, with their strong allegiance to Moses. This may have been his reason for writing P in the first place. Maybe he was so unimpressed that he wrote a torah of his own, to reflect his own people's interests. But the interests of his people, as fellow adherents of Aaron, were similar to the J writer's interests, and so to some extent a counter to the E source existed already in the JE document. Nevertheless, the P source did counteract the original E source in many places. For instance, where the JE source regularly quotes, "And Yahweh said unto Moses . . . ," the P source counters with, "And Yahweh said unto Moses and unto Aaron . . ." The miracles in Egypt are performed with Moses' staff in JE, and with Aaron's staff in P. And so on.

With J representing his interests anyway, the P writer probably had an additional reason for writing his religious document other than the impulse to make a retort to E. Although he was writing in the midst of a resurgence in the fortunes of his particular group, the P writer seemed intent on reviving the traditional values of his people. The P laws, for instance, are noticeably old and customary. The opportunity to write probably arose as a consequence of the time: Hezekiah's reign was known as a time of great literary production. The P writer did fashion some protection against the northern Levites who had flowed into Jerusalem and who would set out to capitalize upon their favored treatment in JE. But he did not allow his allegiance or biases to distort what appears to have been his self-appointed task—to restore appreciation of the divine messages received by Abraham and Moses, on which the ancient Israelite society had been built. God had chosen the ancient Israelites, now the Judeans, as conveyors of his word. This, more than political squabbling, was the primary concern of the writer of P. This, rather than the writer's name, is the information we need to help resolve the Genesis Enigma.

In the P source, with one exception, God is regarded as a transcendent controller of the universe, entirely lacking in anthropomorphic qualities. Jeremiah, and also the prophet Isaiah, had made the comparison of God with the potter, creating people from clay. Fitting with this image, God was depicted as a deity who could change His mind, and change an individual's destiny. This, to me, is a rather primitive plea to the people—a forewarning, or bribe even—to worship God in return for a comfortable life.

Maybe a true holy being, if He exists, would not make such a demand. If God did speak to Abraham and Moses, He would not have made himself known as this type of deity. The possibility exists that the P writer was recording what *really* happened in the days of Moses.

Indeed, much of the Old Testament does not rely on a God that can change the course of events in an individual's life. Biblical laws are comparable to legal laws, or moral responsibilities, today. The P writer depicted a God unlike anything humans could possibly imagine—yet who existed all the same. In P, God is viewed as a distant, transcendent deity, less personal than in J and E, and sometimes harsh. Not all of his doings—not even his very presence on earth—result in enhanced welfare for humans or protection of their interests.

On a similar note, P rejects the concepts of angels and talking animals, such as the talking snake in Eden. In the flood story of P, it does not just rain; there is a cosmic crisis, in which the windows of the heavens and the fountains of the deep break up and the waters that surround our habitable bubble pour in. The JE version of the same ancient Israelite story is that . . . it rains.

The P writer does add some details that reflect his allegiance, such as his stories of sacrifice, which occur only after the consecration of Aaron. But he does not waver far from a very determined stance to reveal the truth without a sugared pill—the truth of a corporeal, material world on to which one could even transpose the physical laws of the universe that we understand today. Personally, *this* made me think further about religion. Maybe we can accept the deductions of modern science and still believe in a God that is beyond science. It could be that the P writer was reaching a similar conclusion—and was reporting the truth that derived from a prophet's message from God. It is hardly surprising that the ostentatious artistic license shown in J and E compelled the P writer to set the record straight.

Yet even where the JE document appeared unadventurous the P writer found it necessary to make changes, probably to remain consistent with the original prophetic messages. A classic case that is of relevance to us occurs in the creation story. The JE document begins with, "In the day that Yahweh God made *earth and heavens* . . ." The P source, on the other hand, states, "In the beginning, God created *the heavens and the earth* . . ." It's a subtle difference, but one that has enormous consequences, as this book has revealed. Here, it is enough to show that the P writer was intent on accurately recording the story of Moses, as it could be useful to future generations. At least, this appears a satisfactory conclusion, or why else would the P writer appear so unyielding?

It should have become clear in the preceding chapters of this book that the P writer would have required considerable biological knowledge, at the very least, to have invented his creation story himself, out of nothing. He lived in Jerusalem, which, as I explained, was expanding rapidly at his time. Jerusalem is positioned toward the southern end of the Jerusalem Saddle, a region between the Hebron Hills in the south and the Bethel Hills in the north. The Jerusalem Saddle has broad valleys and a plain. Jerusalem itself sits 2,400 feet (730 meters) above sea level. On their other side, the Hebron Hills fall away into the Negev, the site of Woolley's and Lawrence's expeditions. In fact the impression of a rather unassuming, meager, insipid fauna that came across in Woolley and Lawrence's description of the Negev can be extended into the Jerusalem Saddle—this is not an area to bring out the biologist in anyone. Most of Jerusalem's rain falls within just three months of the year, which gives rise to barren, parched landscapes, along with the famines noted in the Bible itself. Such environments do not represent the world's diversity of wildlife. Quite the opposite. Then, as we have noted already, the ancient Israelites were not zoologically inclined at the best of times. It would have been extraordinary if the P writer's environment inspired him suddenly to emerge as a zoologist amid a people noted for their disregard for biological, even scientific, inquiry. The prototype "scientists" in history would not appear for a few hundred years yet. The P writer—and, for precisely the same reasons, Moses and Abraham, too—is likely to have lacked the biological knowledge needed to invent his creation story himself.

Indeed, the creation story of P is so concise that it reads like a list, as if an attempt had been made to record a speech, with only the highlights making it onto the page. There is no fleshing out. Elsewhere, P does make one reference to humans created in God's own image, indicating that, although he is rather cautious in comparison with the J and E writers, he still cannot think completely outside the box. He fails to depict God entirely as something beyond the human imagination. But the first page of Genesis does paint such a picture. That first page appears to have been reported as Moses told it.

All in all, the P author's identity and outlook do *not* make his reason for writing the creation account in Genesis 1 appear self-evident. There are no clues associated with his cultural environment to suggest why he wrote that first page of the Pentateuch in the way that he did. Remember that during Woolley and Lawrence's foray into the Bible land, neither they nor the local Bedouin people were moved by or even the least bit interested in whatever wildlife shared their land. Natural history was not their disposition. So did the P author copy his creation account from somewhere else, other than the traditional prose of Moses?

Recent scholarly work on the biblical text has revealed parallels between stories in Genesis and those from the ancient Sumerian and Babylonian worlds, long before the time of Moses. A classic case is the flood story. This suggests that the origin of such stories may predate the J and P sources. But, to add to the trend with which we are becoming familiar, the creation account on the first page of Genesis is an exception. There are no earlier creation stories to compare with the creation account in Genesis 1. The first time such a creation account was recorded was in the P source—it belongs to the ancient Israelites. Indeed, it does appear to have derived from the mouth of Moses, and, as the story goes, to represent the word of God.

The P source was written as an alternative to JE. The writers of J and E were skilled storytellers. In other words, they could exaggerate a story line to enhance its charm. The writer of P clearly felt, probably as a result of such superfluities and the associated tendency toward anthropomorphism, that the Israelites' traditional story of the creation—that derived from Moses—had been misrepresented in JE. The faithful story of the creation—that received by Moses—

had probably endured the generations, and the more doctrinaire writer of the P source felt compelled to record it authoritatively and correct the oversight of the J and E writers. One thing is clear: The writer of P was not in the habit of mindlessly stating the opposite of the contents of the JE document in order to annul its message. Maybe the redactor felt similarly haunted by the misrepresentations of JE when he promoted the traditional Israelite creation account to the beginning of the Pentateuch. In other words, he knew Moses' version too—the people must have kept Moses' prose alive. Otherwise, it would appear strange that the remainder of the Pentateuch is packed with tales that are critically acclaimed today, yet the redactor chose to leave the first page as a collection of short, erratic sentences without praiseworthy literary quality. Maybe an examination of the redactor himself will lead us out of the dark.

THE REDACTOR

The redactor must have possessed a talent for both literature and politics. He put together different opinions and stories packed with different biases and views of ancient Israel through very different eyes. Yet he achieved the near-impossible—he found the perfect balance. He put together the Pentateuch in such a way that all ancient Israelites were satisfied and accepted it as true. Further, they all saw their God in this work, and began to live their lives by it. And these are people with different, often conflicting outlooks. The redactor achieved the equivalent of uniting two disparate political parties today. His was a work of genius.

Whether an additional element of belief existed over two millennia ago—whether the ancient Israelites knew something about the divine that we don't—well, we will never know. Maybe Moses' precise creation story had been kept alive through the generations.

Certainly, it is worth taking a glance at the life and times of the redactor, and the history that led to his seminal work.

Both archaeology and the Bible itself tell us much less about the decades immediately following the destruction of the Temple in Jerusalem in 587 BCE than they do about those that precede it. It appears that Judah's old neighbor Edom seized the moment and encroached on Judah's territory. Meanwhile the Samaritans occupied the northern territory that had been Israel. But we don't know how many Judeans were able to remain in Judah—though we do know that most were in exile.

The Judeans' religion was not compatible with any other, specifically not with the pagan religions of Babylonia and surrounding kingdoms. Yahweh was not a force of nature in the mold of the pagan gods, but existed outside the natural realm, controlling its forces. The Judeans held on to their religion, and so kept their identity alive. But, whether in Judah, Babylonia, or Egypt, they were in somber mood, as several books of the Old Testament attest.

The Babylonians invaded Egypt just nineteen years after the Judeans had fled there. There was no escaping the armies from Babylonia, but the Judeans kept their faith in the more powerful Yahweh by blaming themselves for their predicament—they were not paying for a weakness in their god, but had obviously failed to keep their covenant with Yahweh. Now they needed a place to worship, a center for priests to make sacrifices, and a leader. In 538 BCE they got their break—the Persians, under Cyrus the Great, conquered the Babylonians, and the Judeans were allowed to go home. Cyrus allowed the Judeans to rebuild their homeland and their Temple.

The precious implements of the original, "First" Temple, stolen by the Babylonians, were returned—all except the Ark of the Covenant, the whereabouts of which, as I mentioned, remains a mystery. A new Temple—the Second Temple—was built in Jerusalem to house the First Temple's sacred articles. The Second Temple's Holy of Holies was an empty room, lacking even the golden statues of sphinxes whose wings spread over the ark. This is important to our search for the redactor. And there's another clue tied up in the history of Judea.

Much of the history of this period is obscure. Prophecy diminished and the prophets disappeared. Another mystery is the composition of the new population of Judea—specifically, how many had returned from Babylonia, Egypt, and the former Israel, and how many had simply remained in Judea all along. The figures given in different books of the Bible do not agree.

We do know that the Second Temple was completed in 516 BCE. Many saw this as a fulfillment of a prophecy of Jeremiah. The Second Temple's high priest was an Aaronid—indeed, the entire Temple priesthood was Aaronid at this time: All other Levites were not recognized as legitimate priests, and became assistants to the Aaronids. The Aaronids had successfully reasserted their old claim that they were the only legitimate priests and had completely overcome the priestly advocates of Moses. This may have been partly because the Aaronids had been anti-Babylonian, and as such were favored by the Persians, the current rulers. It may also have been through the influence and power of Ezra.

Only two men in the Bible are known as lawgivers—Moses and Ezra. Ezra was a priest and a scribe. He was an Aaronid priest and his literary skills are associated with one document in particular—"the torah of Moses."

If we follow the system of Richard Elliott Friedman, who as I mentioned previously has sensibly drawn together various strands of the modern approach to biblical authorship, Ezra came from Babylon to Jerusalem in 458 BCE. He came bearing two documents—a letter from the Persian emperor (the most powerful man in the world) giving him authority in Judah, and "the torah of Moses." But what was this "torah of Moses?" The biblical books of Ezra and Nehemiah contain material from JE, D, and P, and so it was likely that this "torah of Moses" was the full Torah—the Pentateuch.

The Persian emperor had appointed a governor of Judah, Nehemiah, who joined forces with Ezra to rebuild the city walls of Jerusalem. The pair also enforced both the observance of the Sabbath and the dissolution of intermarriages between Judeans and others. In the absence of a monarchy, Nehemiah and Ezra led the people of Judah at the beginning of what became known as the Second Temple period, an era that ended with the Temple's destruction by the Romans in 70 BCE.

Ezra first read the Pentateuch to the people at a public assembly at the water gate of Jerusalem. The Judeans—the ancient Israelites—finally had a substantial religious book that some have labeled the First Bible. Because of the turmoil through which they had recently lived, the people treasured this book from the beginning. It was a foundation on which to rebuild their culture and society. It gave them a sense of heritage from a remarkable past. The last question remaining in this chapter is: Who put together the Pentateuch from its component parts?

The redactor's work was extraordinary. He began with three documents reflecting different approaches and allegiances—JE, D, and P—and finished with one book that satisfied everyone. The chasms he had to close were vast, but each document was so well known to the people that they all had to stay. JE championed Moses, and held that any Levite could be a priest. P championed Aaron, and instructed that only men descended from Aaron could be priests. JE contained angels, talking animals, and an anthropomorphic God. P must have felt that the J and E writers had strayed too far from the initial messages they were documenting, and that the people would digest the original divine communication without the need for fantasy. To make matters worse, the interests of the P writer further clashed with those of the D writer in another Aaron-Moses rift.

Clearly, the redactor could not leave the four documents as complete, separate books standing alongside one another; so he cut each document into pieces and wove the pieces together in an intricate fabric. The finished work—the Pentateuch—was so accomplished that no one even noticed it had once been four disparate works, at least not for a couple of thousand years anyway.

The redactor had used the P source as the framework for the first four books, and the D source for the last. The D source did contain a complete farewell speech for Moses, and so this fitted well with a book about Moses' life and times. Then, the P source was the largest and most comprehensive, and so provided a skeleton for JE and D to flesh out, while giving continuity to the stories. Another explanation, of course, is that the redactor was favoring the P source. It appears that, like the P author, the redactor was an Aaronid priest, and as such a male.

The redactor inserted some texts of his own that explicitly express a priest's concerns and interests, and favor Aaron, while using the P source's linguistic style. In addition, the priests were the group with access to documents. But the redactor's own text not only confirms his priestly status; it also indicates the moment in time when someone in a long line of priests—the guardians of J, E, P, and D—chose to combine them.

The redactor wrote chapter 15 of the book of Numbers. It concerns sacrifice, and covers the same material already found in P, except that it never mentions the Tabernacle—the building imperative to sacrifice in P. This is a speaking omission, indicating that the Tabernacle no longer stood at the time of writing. The redactor must have worked after 587 BCE. Yet sacrifices still take place in the redactor's text—at a temple with no Tabernacle. Clearly, the Second Temple existed in the redactor's time. He must have written after 516 BCE.

The Second Temple period was without kings, and the Aaronid priests were in charge. It is not surprising that this was the moment when the priests were first able to combine the four sacred texts and bestow complete authority on the new book. But who, precisely, was the redactor? One individual stands out in the history of Judah as an Aaronid priest in the Second Temple days with unique power and the conviction to exercise it. That priest was Ezra.

Ezra had the status to instruct the people on what their authoritative religious book should be. He had the backing of the emperor and the necessary powers of enforcement. As I mentioned, only two men in the entire Bible were known as lawgivers—Moses and Ezra. But Ezra was also a scribe.

The Bible spells out that Ezra "had set his heart on seeking out Yahweh's Torah." The emperor authorizes him to teach and enforce "the law of your God which is in your hand." Further, it says that Ezra "was a ready scribe in the Torah of Moses." We know that Ezra was not involved in the writing of J, E, D, or P. So the Bible itself is telling us that Ezra was the redactor. Ezra gave the Pentateuch its first public reading because *he* had put it together. Moses had become known as the author of the four separate sources, so Ezra combined them as to produce a story flowing from beginning to end, all seemingly from Moses' quill. This the people could digest.

Ezra's artistry and creativity formed the final link in preserving the ancient Israelites' early history and prophetic messages. It is uncertain whether he shared the P writer's possible discomfort at the JE misrepresentation of the creation story. Nor are there any clues to suggest he *knew* that the creation account in P was the very words of Moses—those apparently from God. Yet from what we know of Ezra's history and intentions, there appears no personal reason why he would wish to favor this account of the history of the earth and of life. There is nothing in this account from which Ezra—or the P writer, for that matter—could have benefited in any way. It is likely that this creation account was special to the people—people to whom prophets and strict religious principle were highly important, and people who accepted only truth. Ezra was producing a story of Moses, to appear as if written by Moses himself. What better way to start, and to draw the people, than by actually using Moses' own words?

So our glimpse into the identity of the redactor only confirms our suspicions that the creation account of Genesis 1 had, to the best of the ancient Israelites' knowledge, divine origins in that it came from Moses. The redactor was a literary man and not the type to let a false story slip through his fine-meshed net. He would have been even less likely to allow anything remotely doubtful to occupy his *first* page. On the contrary, he would have given most thought to that page, for which he would have been most strictly judged and held most accountable. He had to have been absolutely certain that this was God's version of the *big* event—the true version of how the earth, the stars, and humans came to be here. Certainly, there appears nothing in the character or the environment of the P writer, the redactor, or even Moses himself to suggest how or why any of them would invent an account such as this.

The end result of the Pentateuch's formation process was that all the writers involved captured something in their text, something that was not lost on the redactor. Whatever this was, *it engaged the ancient Israelites and people of all generations thereafter*. The Old Testament remains the commonest book on our bookshelves today, over two thousand years since its formation. All those other religions that existed in tandem with the ancient Israelite religion—the Abrahamic religion—quickly fell by the wayside. The Abrahamic religion, on the other hand, grew to be dominant today. The Pentateuch is central to the modern Christian, Jewish, and to some extent Islamic religions. Maybe there really is something special about it . . . or maybe it is just a history book.

BIBLICAL AUTHORSHIP AND THE GENESIS ENIGMA

In this appendix we have learned that it was an Aaronid priest who wrote the Pentateuch's P source, including the creation account, and that this account was true to the history of his people—the ancient Israelites. But, more important, we have also learned something of the character of the people involved in the Old Testament's formation. They did not represent the origins of civilization—far from it—and they lived at a time when writing was a well-developed practice, even fashionable. Most important, they were not scientists. They predated science, and the way they thought was not conducive to the scientific method.

That Aaronid priest who wrote the Bible's first page, or Moses, who may have given the ancient Israelites their creation account originally, lacked any interest in natural history. Although demonstrating attention to detail in other subjects—geography, politics, economics, law—this Aaronid priest, and the character Moses, provide us with no signs of a biological inclination. The scientific method, necessary to decipher the true account of how the universe formed and life evolved, with its repeatable experiments, was yet to manifest itself. The ancient Israelites were not conducting scientific experiments in their sheds—if they were, they would have written about it, as they wrote about everything else they did. The writer of the Bible's first page simply roamed the desert or traversed the dusty streets of ancient Jerusalem during the day, and marveled at the stars at night. He was without so much as a magnifying lens.

Indeed, the history books tell us that science and natural history began some centuries later with the ancient Greeks, who were influenced by very different natural surroundings. So, in terms of providing an explanation for how the universe and life came to be, the Aaronid priest given this task, or the character Moses, would not have had a clue. All the same, something was written. And it made its way to pride of place in the Bible.

As such, unprovided with evidence of any kind, the creation account on the Bible's opening page might be assumed a fantasy. But the Genesis Enigma has told us that those enigmatic phrases that ignite the Bible actually mean something—they are scientifically accurate. That would be an outrageous assertion, were it not true. The conclusion is that *this page of the Bible could, perhaps more than any other, represent God's hand in the Bible*. The true account of how we came to exist may have been handed to humans by God.

In any case, our strong preconception that science has, with each discovery, chipped away at the notion of God is proved wrong in this book. Now we can live with the real possibility that God exists while fully accepting science, rather than straining to find contradictions. Faith suddenly appears that much stronger.

REFERENCES

Introduction

The Holy Bible: Today's New International Version. London: Hodder & Stoughton, 2004.

King James Folio Bible. 2nd ed. London: Robert Barker, 1613.

King, R. *Michelangelo and the Pope's Ceiling*. London: Chatto & Windus, 2002.

Mancinelli, F., and L. Heusinger. *The Sistine Chapel*. London: Constable & Co., 1978.

Mancinelli, F. "The Technique of Michelangelo as a Painter: A Note on the Cleaning of the First Lunettes in the Sistine Chapel." *Apollo* (May 1983): 362–367.

Michelangelo Buonarroti. *Michelangelo: A Record of His Life as Told in His Own Letters and Papers*. Translated and edited by R. W. Carden. London: Constable & Co., 1913.

Chapter 1

Blair, J. *The Chronology and History of the World: From the Creation to the Year of Christ*. London, 1779.

Bowman, S., ed. *Science and the Past*. London: British Museum Press, 1991.

Coogan, M. D. *The Old Testament: A Historical and Literary Introduction to the Hebrew Scriptures*. New York: Oxford University Press, 2006.

Davis, T. W. *Shifting Sands: The Rise and Fall of Biblical Archaeology*. New York: Oxford University Press, 2004.

Evetts, B.T.A. *New Light on the Bible and the Holy Land: Being an Account of Some Recent Discoveries in the East*. New York: Cassell, 1894.

Fernández-Armesto, F. *Truth: A History and a Guide for the Perplexed*. London: Bantam Press, 1997.

Finkelstein, I., and N. A. Silberman. *The Bible Unearthed: Archaeology's New Vision of Ancient Israel and the Origin of Its Sacred Texts*. New York: Touchstone, 2002.

Frye, N. *The Great Code: The Bible and Literature*. First published 1982. Edited with an introduction by Alvin A. Lee. Toronto: University of Toronto Press, 2006.

Gosse, P. H. *The Ancient and Modern History of the Rivers of the Bible*. 2nd ed. London: G. Cox, 1850.

Hilprecht, H. V. *Explorations in Bible Lands during the 19th Century*. Edinburgh: T. & T. Clark, 1903.

Hogarth, D. G. *Carchemish: Report on the Excavations at Djerabis on Behalf of the British Museum, Parts I and II*. First published 1914. London: British Museum Press, 1969.

Knightley, P., and C. Simpson. *The Secret Lives of Lawrence of Arabia*. London: Panther, 1971.

Lawrence, T. E. *The Diary Kept by T. E. Lawrence While Travelling in Arabia During 1911*. London: Corvinus Press, 1937. Reprinted by Hobbs, Southampton.

Lindsay, Lord. *Letters on Egypt, Edom, and the Holy Land*. 4th ed. London: Henry Colburn, 1847.

Mitchell, T. C. *The Bible in the British Museum: Interpreting the Evidence*. London: British Museum Press, 2004.

Moscrop, J. J. *Measuring Jerusalem: The Palestine Exploration Fund and British Interests in the Holy Land*. London: Leicester University Press, 1999.

Prideaux, H. *The Old and New Testament Connected in the History of the Jews and Neighbouring Nations from the Declension of the Kingdoms of Israel and Judah to the Time of Christ*. Dublin: A. Rhames, 1719.

Rihani, A. *Around the Coasts of Arabia*. London: Constable & Co., 1930.

Rowley, H. H. *Growth of the Old Testament*. London: Hutchinson's University Library, 1950.

Winstone, H. V. F. *Woolley of Ur: The Life of Sir Leonard Woolley*. London: Secker & Warburg, 1990.

Woolley, C. L. *Digging Up the Past*. 3rd ed. Tonbridge: Benn, 1973.

Woolley, C. L., and T. E. Lawrence. *The Wilderness of Zin*. London: Jonathan Cape, 1936.

CHAPTER 2

Aristotle. *On Man in the Universe*. Edited by L. R. Loomis. New York: Gramercy Books, 1971.

Barrow, J. D. *The Origin of the Universe*. London: Weidenfeld & Nicolson, 1994.

Benton, M. J., ed. *The Seventy Great Mysteries of the Natural World*. London and New York: Thames & Hudson, 2008.

Blackburn, S. *Plato's Republic: A Biography*. New York: Atlantic Monthly Press, 2007.

Copernicus, N. *On the Revolutions*. Translated and with commentary by E. Rosen. In *Nicholas Copernicus: Complete Works*, vol. 2. Baltimore: Johns Hopkins University Press, 1992.

Grant, E. *Science and Religion, 400 B.C. to A.D. 1550: From Aristotle to Copernicus*. Westport, Conn.: Greenwood Press, 2004.

Hooykaas, R. *G. J. Rheticus' Treatise on Holy Scripture and the Motion of the Earth*. With translation, annotations, commentary and additional chapters on Ramus-Rheticus, and the development of the problem before 1650. Amsterdam: Elsevier, 1984.

Hubble, E. *The Realm of the Nebulae*. New Haven: Yale University Press, 1936.

Lindberg, D. C., and R. L. Numbers, eds. *God & Nature: Historical Essays on the Encounter between Christianity and Science*. Berkeley: University of California Press, 1986.

Newton, I. *Philosophiae Naturalis Principia Mathematica*. 3rd ed. London: W. & J. Innys, 1726.

Nordmann, C. *Einstein and the Universe: A Popular Exposition of the Famous Theory*. Translated by J. McCabe. London: T. Fisher Unwin, 1922.

Olson, R. G. *Science and Religion, 1450–1900: From Copernicus to Darwin*. Baltimore: Johns Hopkins University Press, 2006.

Regiomontanus, J. *On Triangles [De Triangulis Omnimodis]*. Edited by J. Schöner, translated by B. Hughes. Madison: University of Wisconsin Press, 1967.

Repcheck, J. *Copernicus' Secret: How the Scientific Revolution Began*. New York: Simon & Schuster, 2007.

CHAPTER 3

Buckland, W. *Reliquiae Diluvianae; or, Observations on the Organic Remains Contained in Caves, Fissures, and Diluvial Gravel, and on other Geological Phenomena, Attesting the Action of an Universal Deluge*. London: John Murray, 1824.

Chambers, R. *Vestiges of the Natural History of Creation and Other Evolutionary Writings*. Copy of the 1844 First Edition. Edited with a new introduction by J. A. Secord. Chicago: University of Chicago Press, 1994.

Descartes, R. *The Philosophical Works of Descartes*. Edited by E. Haldane and G. R. T. Ross. London: Cambridge University Press, 1967.

Geikie, A. *Life of Sir Roderick I. Murchison*. 2 vols. London: John Murray, 1875.

Glennie, K. W. *The Desert of Southeast Arabia*. Manama, Bahrain: Gulf PetroLink, 2005.

Hutton, J. "The Theory of the Earth." *Transactions of the Royal Society of Edinburgh* 1 (1788): 209–304.

Lyell, C. *Principles of Geology*. 3 vols. London: John Murray, 1831–1833.

Owen, R. *The Life of Richard Owen by His Grandson*. 2 vols. London: John Murray, 1894.

Picard, L. *Geological Researches in the Judean Desert*. Jerusalem: Goldberg's Press, 1931.

Thompson, K. *Before Darwin: Reconciling God and Nature*. New Haven: Yale University Press, 2005.

CHAPTER 4

Bates, H. W. *The Naturalist on the River Amazons*. 3rd ed. London: John Murray, 1875.

Broberg, G. *Carl Linnaeus*. Translated by R. Tanner. Stockholm: The Swedish Institute, 1992.

Darwin, C. *On the Origin of Species by Means of Natural Selection*. 6th ed. London: John Murray, 1872.

Darwin, E. *The Temple of Nature; or, the Origin of Society: A Poem.* London: T. Bensley, 1803.

Dulbecco, R. *The Design of Life.* New Haven: Yale University Press, 1987.

Gosse, P. H. *Omphalos: An Attempt to Untie the Geological Knot.* London: John van Voorst, 1857.

Gosse, P. H. *A History of the British Sea-Anemones and Corals.* London: John van Voorst, 1860.

Hochachka, P. W., and G. N. Somero. *Biochemical Adaptation.* Princeton, N.J.: Princeton University Press, 1984.

Huxley, T. H. *Darwiniana: Collected Essays.* 2 vols. London: Macmillan, 1893.

Johnson, A. P., H. J. Cleaves, J. P. Dworkin, D. P. Glavin, A. Lazcano, and J. L. Bada. "The Miller Volcanic Spark Discharge Experiment." *Science* 322, no. 5900 (2008): 404.

Knoll, A. H. *Life on a Young Planet: The First Three Billion Years of Life on Earth.* Princeton, N.J.: Princeton University Press, 2004.

Linnaeus, C. *Systema Naturae.* 7th ed. Leipzig, Germany: G. Kiesewetteri, 1748.

Miller, S. L., and L. E. Orgel. *Origins of Life on the Earth.* Englewood Cliffs, N.J.: Prentice Hall, 1974.

Oparin, A. I. *Origin of Life.* 2nd ed. New York: Dover, 1953.

Paley, W. *Natural Theology or Evidences of the Existence and Attributes of the Deity Collected from the Appearance of Nature.* London: Hallowell, 1802.

Ridley, M., ed. *Evolution.* 2nd ed. Oxford and New York: Oxford University Press, 2004.

Thwaite, A. *Glimpses of the Wonderful: The Life of Philip Henry Gosse.* London: Faber and Faber, 2002.

Wallace, A. R. *The Malay Archipelago: A Narrative of Travel, with Studies of Man and Nature.* 3rd ed. London: Macmillan, 1872.

Wallace, A. R. *Darwinism: An Exposition of the Theory of Natural Selection with Some of Its Applications*. London: Macmillan, 1889.

Wallace, A. R. *Natural Selection and Tropical Nature: Essays on Descriptive and Theoretical Biology*. New ed. London: Macmillan, 1891.

CHAPTER 5

Briggs, D.E.G., and R. A. Fortey. "Wonderful Strife: Systematics, Stem Groups, and the Phylogenetic Signal of the Cambrian Radiation." *Paleobiology* 31, no. 2 (2005): 94–112.

Buffon, G.-L. *Le Buffon des Enfans, Ou Petite Histoire Naturelle*. 3rd ed. Paris: Belin, 1816.

Clarkson, E., R. Levi-Setti, and G. Horváth. "The Eyes of Trilobites: The Oldest Preserved Visual System." *Arthropod Structure and Development* 35 (2006): 247–259.

Coultas, H. *Zoology of the Bible*. London: Wesleyan Conference Office, 1876.

Crick, F.H.C. *Of Molecules and Men*. 1966. Amherst, N.Y.: Prometheus, 2004.

Cuvier, G. *The Animal Kingdom*. New ed. London: Henry G. Bohn, 1863.

Fisher, R. A. *The Genetical Theory of Natural Selection*. Facsimile of the Original 1930 Edition. Oxford: Oxford University Press, 2003.

Fortey, R. *Trilobite! Eyewitness to Evolution*. London: HarperCollins, 2000.

Glaessner, M. F. *The Dawn of Animal Life: A Biohistorical Study*. Cambridge and New York: Cambridge University Press, 1984.

Hillel, D. *The Natural History of the Bible: An Environmental Exploration of the Hebrew Scriptures*. New York: Columbia University Press, 2006.

Hooke, R. *Micrographia, or, Some Physiological Descriptions of Minute Bodies Made by Magnifying Glasses, with Observations and Inquiries Thereupon*. London: J. Martyn and J. Allestry, 1665.

Jenkins, R.J.F. "Aspects of the Geological Setting and Palaeobiology of the Ediacara Assemblage." In M. Davies et al., eds., *Natural History of the Flinders Ranges*, pp. 33–45. Adelaide, Australia: Royal Society of South Australia, 1996.

Land, M. F., and D.-E. Nilsson. *Animal Eyes*. Oxford: Oxford University Press, 2001.

Marshall, C. R. "Explaining the Cambrian 'Explosion' of Animals." *Annual Review of Earth and Planetary Sciences* 34 (2006): 355–384.

McMenamin, M.A.S. *The Garden of Ediacara: Discovering the First Complex Life*. New York: Columbia University Press, 1998.

Mendel, G. J. (1865) *Versuche über Pflanzen-Hybriden. Verh. Naturf. Ver. in Brünn,* Volume iv. For an English reference, see R. M. Henig, *Monk in the Garden: The Lost and Found Genius of Gregor Mendel, the Father of Genetics.* New York: Houghton Mifflin, 2000.

Murray, A. *The Geographical Distribution of Mammals*. London: Day and Son, 1866.

Nilsson, D.-E., and S. Pelger. "A Pessimistic Estimate of the Time Required for an Eye to Evolve." *Proceedings of the Royal Society, London B* [Biological Sciences] 256, no. 1345 (1994): 53–58.

Parker, A. R. "Colour in Burgess Shale Animals and the Effect of Light on Evolution in the Cambrian." *Proceedings of the Royal Society, London B* [Biological Sciences] 265, no. 1400 (1998): 967–972.

Parker, A. R. *In the Blink of an Eye*. London: Simon & Schuster, and Cambridge, Mass.: Basic Books, 2003.

Parker, A. R. "On the Origin of Optics." *Optics & Laser Technology* 41 (2009).

Tudge, C. *The Variety of Life: A Survey and a Celebration of All the Creatures That Have Ever Lived*. Oxford and New York: Oxford University Press, 2000.

Wallace, A. R. *The Geographical Distribution of Animals*. 2 vols. London: Macmillan, 1876.

CHAPTER 6

Briggs, D.E.G., D. H. Erwin, and F. J. Collier. *The Fossils of the Burgess Shale*. Washington, D.C.: Smithsonian Institution Press, 1994.

Conway Morris, S. "The Community Structure of the Middle Cambrian Phyllopod Bed (Burgess Shale)." *Palaeontology* 29 (1986): 423–467.

Gould, S. J. *Wonderful Life: The Burgess Shale and the Nature of History*. New York: W. W. Norton, 1989.

Lawrence, T. E. "The Evolution of a Revolt." *Army Quarterly and Defence Journal* 1 (1920): 55–69.

Neilsen, C. *Animal Evolution: Interrelationships of the Living Phyla*. Oxford and New York: Oxford University Press, 2001.

Parker, A. R. "Discovery of Functional Iridescence and Its Coevolution with Eyes in the Phylogeny of Ostracoda (Crustacea)." *Proceedings of the Royal Society, London B* [Biological Sciences] 262, no. 1365 (1995): 349–355.

Parker, A. R. "Light-Reflection Strategies." *American Scientist* 87, no. 3 (1999): 248–255.

Parker, A. R. "515 Million Years of Structural Colour." *Journal of Optics A: Pure and Applied Optics* 2 (2000): R15–R28.

Walcott, C. D. "Abrupt Appearance of the Cambrian Fauna on the North American Continent: Cambrian Geology and Paleontology II." *Smithsonian Miscellaneous Collections* 57 (1910): 1–16.

Whittington, H. B. "The Burgess Shale: History of Research and Preservation of Fossils." In *Proceedings of the North American Paleontological Convention, Chicago, 1969*, vol. 1: 1170–1201. Lawrence, Kans.: Allen Press, 1971.

Xian-guang, H., R. J. Aldridge, J. Bergstrom, D. J. Siveter, and F. Xiang-hong. *The Cambrian Fossils of Chengjiang, China*. Malden, Mass., and Oxford: Blackwell, 2004.

Zhu, M.-Y., L. E. Babcock, and S.-C. Peng. "Advances in Cambrian Stratigraphy and Paleontology: Integrating Correlation Techniques, Paleobiology, Taphonomy and Paleoenvironmental Reconstruction." *Palaeoworld* 15 (2006): 217–222.

CHAPTER 7

Clack, J. A. *Gaining Ground: The Origin and Evolution of Tetrapods*. Bloomington, Ind.: Indiana University Press, 2002.

Gilmour, I., and M. A. Sephton. *An Introduction to Astrobiology*. Cambridge: Cambridge University Press, 2004.

Knight, C. R. *Life Through the Ages.* Bloomington, Ind.: Indiana University Press, 2001.

Murchison, R. I. *Siluria*. 5th ed. London: John Murray, 1872.

Owen, R. *A History of British Fossil Mammals and Birds*. London: John van Voorst, 1846.

Rudwick, M.J.S. *The Great Devonian Controversy: The Shaping of Scientific Knowledge among Gentlemanly Specialists*. Chicago: University of Chicago Press, 1985.

Southwood, R. *The Story of Life*. Oxford and New York: Oxford University Press, 2003.

Wallace, A. R. *Australasia*. Part of *Stanford's Compendium of Geography and Travel* (1878); based on Hellwald's *Die Erde und Ihre Völker*. London: Edward Stanford, 1879.

CHAPTER 8

Buller, W. L. "Notes on New Zealand Birds." *Transactions and Proceedings of the Royal Society of New Zealand* 25 (1892): 53–63.

Gould, J. *Hummingbirds*. Reprinted from the original 1861 edition. London: Wellfleet, 1990.

Greenway, J. C. *Extinct and Vanishing Birds of the World*. Special Publication No. 13. New York: American Committee for International Wild Life Protection, 1958.

Knight, C. *Pictorial Museum of Animated Nature*. London: London Printing and Publishing Co., ca. 1870.

Mueller, T. "Biomimetics: Design by Nature." *National Geographic* (April 2008): 68–91.

Parker, A. R., R. C. McPhedran, D. R. McKenzie, L. C. Botten, and N.-A. P. Nicorovici. "Photonic Engineering: Aphrodite's Iridescence." *Nature* 409 (2001): 36–37.

Parker, A. R., V. L. Welch, D. Driver, and N. Martini. "Structural Colour: Opal Analogue Discovered in a Weevil." *Nature* 426 (2003): 786–787.

Parker, A. R. *Seven Deadly Colours*. London: Simon & Schuster, 2005.

Parker, A. R., and H. E. Townley. "Biomimetics of Photonic Nanostructures." *Nature Nanotechnology* 2 (2007): 347–353.

Rhodes, R. *John James Audubon: The Making of an American*. New York: Alfred A. Knopf, 2004.

Swinton, W. E. *Fossil Birds*. London: British Museum (Natural History), 1958.

Tree, I. *The Bird Man: A Biography of John Gould*. London: Ebury Press, 2003.

CHAPTER 9

Cornwell, J. *Darwin's Angel: An Angelic Riposte to* The God Delusion. London: Profile, 2007.

Dawkins, R. *The God Delusion*. London: Bantam, 2006.

Digby, K. *Religio Medici*. 8th ed. Mentson, U.K.: Scolar Press, 1973.

Doyle, A. C. *The New Revelation*. London: Hodder and Stoughton, 1918.

Lennox, J. C. *God's Undertaker: Has Science Buried God?* Oxford: Lion, 2007.

Lewis, C. S. *Mere Christianity*. First published 1943. New York: Touchstone, 1996.

Locke, J. *An Essay Concerning Human Understanding*. Edited by P. Nidditch. Oxford: Oxford University Press, 1975.

McGrath, A. E. *Dawkins' God: Genes, Memes, and the Meaning of Life*. Oxford: Blackwell, 2005.

Nichols, P. *Evolution's Captain: The Tragic Fate of Robert Fitzroy, the Man Who Sailed Charles Darwin Around the World*. London: Profile, 2003.

Pew Research Center: http://people-press.org/commentary/?analysis id=118.

Russell, B. *A History of Western Philosophy*. London: Allen and Unwin, 1946.

Salmi, M., L. Becherucci, A. Marabottini, and A. F. Tempesti. *Raffaello: The Paintings, the Drawings*. London: Grange, 1998.

Shorey, P. *Platonism, Ancient and Modern*. Berkeley: University of California Press, 1938.

Spadaro, G. I. *The Esoteric Meaning in Raphael's Paintings: The Philosophy of Composition in The Disputa, The School of Athens, and The Transfiguration*. Great Barrington, Mass.: Lindisfarne Books, 2006.

Wallace, A. R. *Man's Place in the Universe: A Study of the Results of Scientific Research in Relation to the Unity or Plurality of Worlds*. London: Chapman and Hall, 1903.

Wallace, A. R. *The World of Life: A Manifestation of Creative Power, Directive, Mind and Ultimate Purpose*. 3rd ed. London: Chapman and Hall, 1911.

APPENDIX

Buber, M. *Moses*. Oxford: East and West Library, 1947.

Fraser, J. B. *Mesopotamia and Assyria: From the Earliest Ages to the Present Time; with Illustrations of Their Natural History*. 2nd ed. Edinburgh: Oliver & Boyd, 1842.

Friedman, R. E. *Who Wrote the Bible?* San Francisco: HarperSan-Francisco, 1989.

Huxley, J. *From an Antique Land: Ancient and Modern in the Middle East*. New York: Crown, 1956.

Mellaart, J. *Earliest Civilizations of the Near East*. London: Thames and Hudson, 1965.

Phillips, G. *The Moses Legacy: The Evidence of History*. London: Sidgwick & Jackson, 2002.

Rogerson, J. W., and P. R. Davis. *The Old Testament World*. London: T. & T. Clark, 2005.

Woolley, C. L. *Ur of the Chaldees: A Record of Seven Years of Excavation*. London: Ernest Benn, 1931.

ACKNOWLEDGMENTS

The Genesis Enigma began as an idea, which I shared with an ecological acquaintance and then friend, Rosie Boycott. Rosie encouraged me to tell the tale, and introduced me to an agent, Charlie Viney, and everything fell into place. Numerous lunches and coffees with Rosie and Charlie concluded in a coherent, balanced, and above all rational account of a religious enigma that could be important. I'm grateful to Rosie and Charlie.

The Genesis Enigma developed over yet more lunches, this time at Green College (now Green Templeton College), Oxford. In particular, Professor John Lennox orchestrated a series of fascinating meetings with experts in this field, of which he is one. My thanks go to John and his colleagues.

I would also like to thank Felicity Cobbing of the Palestine Exploration Fund for allowing me access to its headquarters in London, and to Jonathan Tubb of the British Museum for introducing me to its biblical collections.

None of this would have been possible without the discussions with, and support and encouragement of my wife and family—thank you all.

Finally, my ideas were hammered into shape by my editors—Simon Thorogood in London and Stephen Morrow in New York, with some valuable advice from Meredith Curnow in Sydney and Gillian Somerscales in Oxford. Simon played a great devil's advocate and further resolved the coherence and balance of my arguments on a finer scale. My last thanks go to you.

INDEX

Note: Page number in *italics* refer to illustrations or maps.